SpringerBriefs in Energy

Energy Analysis

Series Editor

Charles A. S. Hall

AF736961

For further volumes:
http://www.springer.com/series/10041

Roger Boyd

Energy and the Financial System

What Every Economist, Financial Analyst, and Investor Needs to Know

Springer

Roger Boyd
Vice President (Retired), Bank of Montreal
Toronto, ON
Canada

ISSN 2191-5520 ISSN 2191-5539 (electronic)
ISBN 978-3-319-04237-4 ISBN 978-3-319-04238-1 (e-Book)
DOI 10.1007/978-3-319-04238-1
Springer Cham Heidelberg New York Dordrecht London

Library of Congress Control Number: 2014930311

© Roger Boyd 2013
This work is subject to copyright. All rights are reserved by the Publisher, whether the whole or part of the material is concerned, specifically the rights of translation, reprinting, reuse of illustrations, recitation, broadcasting, reproduction on microfilms or in any other physical way, and transmission or information storage and retrieval, electronic adaptation, computer software, or by similar or dissimilar methodology now known or hereafter developed. Exempted from this legal reservation are brief excerpts in connection with reviews or scholarly analysis or material supplied specifically for the purpose of being entered and executed on a computer system, for exclusive use by the purchaser of the work. Duplication of this publication or parts thereof is permitted only under the provisions of the Copyright Law of the Publisher's location, in its current version, and permission for use must always be obtained from Springer. Permissions for use may be obtained through RightsLink at the Copyright Clearance Center. Violations are liable to prosecution under the respective Copyright Law.
The use of general descriptive names, registered names, trademarks, service marks, etc. in this publication does not imply, even in the absence of a specific statement, that such names are exempt from the relevant protective laws and regulations and therefore free for general use.
While the advice and information in this book are believed to be true and accurate at the date of publication, neither the authors nor the editors nor the publisher can accept any legal responsibility for any errors or omissions that may be made. The publisher makes no warranty, express or implied, with respect to the material contained herein.

Printed on acid-free paper

Springer is part of Springer Science+Business Media (www.springer.com)

Contents

About the Author

Roger Boyd is a retired financial industry executive, where he worked for 25 years. He received a BSc in Information Systems from Kingston University in England, an MBA in Finance from Stern School of Business, New York University in the United States, and an MA in Integrated Studies from Athabasca University in Canada. Over the past decade he has taken a deepening interest in the way in which modern societies, especially their financial systems, will deal with gobal threats such as energy constraints and climate change. He also maintains a blog covering such issues, www.humanitystest.com.

Chapter 1
The Nature of the Problem

Over the next decade or so, the most basic assumptions you have about how the economy and the financial system work will, almost certainly, be turned upside down. The perpetual economic growth you have always known will become less and less possible as the cheap energy powering it steadily diminishes. As more energy is required to extract fossil fuels or to provide alternative energy sources there will be less energy for everything else. That 'everything else' includes both your favourite company's earnings and the ability for individuals, companies, and governments to keep paying their debts. The financial system that you assume will multiply your savings and provide for your retirement is completely dependent upon continued growth; and without that growth, it will cease to function. Societies may be able to mitigate these impacts through increased energy efficiency and other measures, but these will only slow down the impacts rather than being a cure for the underlying scarcity of cheap energy.

Rapid economic growth and the modern financial system really took off about 200 years ago with the exploitation of fossil fuels. Many new institutions were required to support and fund this growth, such as shareholder-owned joint stock companies (the predecessors of the modern corporation), corporate bonds, and stock markets [1]. Without these innovative ways of concentrating and directing investment capital, the financial system could not have supported such rapid growth. As that growth started to produce a middle class, and then even a working class, who could afford more than mere subsistence, the consumer finance industry came into being and grew rapidly. From the 1970s onwards, extensive financial deregulation and the delinking of money from gold accelerated the pace of growth. The post-war industrialization of Japan, then South Korea and Taiwan, followed later by the behemoths of Brazil, China, and India provided still more growth opportunities for finance. The opening up of the USSR and Eastern Europe to capitalism further added to these opportunities. After 200 years of this fossil-fuelled growth, the financial system has become completely dependent upon its continuance. However, without continued access to cheap sources of energy, that growth is impossible as the tight correlation between global energy use and economic growth shows (Fig. 1.1).

There are those who say that growth has been decoupled from energy use; but they are wrong. The rich countries have moved their most energy-intensive produc-

R. Boyd, *Energy and the Financial System*, SpringerBriefs in Energy,
DOI 10.1007/978-3-319-04238-1_1, © Roger Boyd 2013

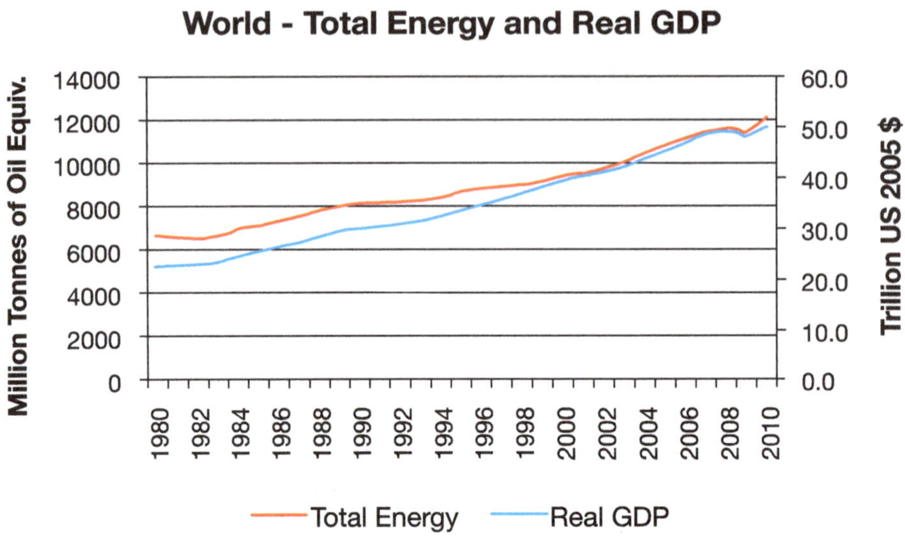

Fig. 1.1 Relationship between global energy usage and global GDP. (Source: Is it really possible to decouple GDP growth and Energy Growth? Gail Tverberg, Our Finite World website article, November 15th, 2011)

tion offshore to places like China and India. The result is an apparent decoupling of energy and growth. However, if the energy used to produce the goods and services actually consumed in the rich countries—including imports—is analyzed, the tight linkage is again apparent. Over the past 200 years, much of humanity has had a wonderful party underwritten by exploiting huge deposits of fossil fuel. Unfortunately, we have managed to burn through these deposits at an astonishing rate with those easiest (and thus cheapest) to extract being used up first. While we are not going to run out of fossil fuels anytime soon, we are running out of the cheap stuff which funded our modern societies, economies, and financial systems.

The cost of extracting fossil fuels can be treated as we would any other investment; there is an amount that has to be invested and an amount that will be returned by that investment. If the return is not high enough then the investment will not be made. In the case of energy, the best insight is gained by looking at the amount of energy invested and the amount of energy returned; so instead of *dollars invested* and *dollars returned* we will use *energy invested* and *energy returned*, to produce an *Energy* Return on Investment (EROI). When the first commercial oil and gas wells were drilled, the EROI was incredible; in the region of 30: 1, and for coal mining about 60:1 [2]. That is like investing $ 1 and getting $ 30–60 in return. These incredible returns are what fuelled the creation and growth of our complex modern economies.

As time passed, our energy demands grew, and some of the easy and cheap oil and gas wells showed the beginnings of depletion. With that 'low hanging fruit' drying up, more difficult, more expensive conventional wells had to be utilized.

The EROI of the ever-newer conventional oil and gas wells had already fallen to about 18 : 1 a few years ago, and is at about 10 : 1 presently [2]. In practical terms, this means that for a given amount of oil or gas produced, the net amount of energy available to society becomes less, as more energy has to be used for exploration and extraction. On average, for the unconventional oil obtained from tar sands, or oil and gas produced from deep-sea wells and by hydraulic fracturing (fracking), the EROI is less than 10 : 1 [2]. For now, because oil is a globally traded 'fungible' commodity, the amount of oil flowing from old, established high EROI fields keeps the average EROI reasonably high. Natural gas supply is more locally traded, which shows up in large regional price differences, but its global EROI is also kept reasonably high by the older, cheaper wells. The use of unconventional resources may maintain production levels, but is a sign that the days of cheap energy are numbered as the global EROI continues to decline. Each 'down tick' of the average EROI for oil and gas will siphon a greater share of the available energy into the extraction process.

As for the non-fossil energy alternatives, solar PV, bio-fuels, and nuclear power also have the same low, or even lower, EROIs than the new or unconventional oil and gas. The widespread attempts at replacing depleting low-cost oil and gas with them only add to the decreasing net energy problem. Currently, only hydro-power and wind provide net energy levels equal to conventional oil and gas. Hydropower is severely constrained by availability of usable sites, and wind power currently provides only 1 % of global energy supplies so even with very rapid growth it will remain a small contributor for a decade or more. That leaves the oldest and dirtiest of the fossil fuels, coal, which still has a high EROI [2]. Society is currently living off coal plus the high EROI oil and gas wells, the vast majority of which are decades old and rapidly depleting. Those old wells, together with coal, and a recession-driven reduction in oil usage by Europe and the USA, are keeping modern society going. In recent years, the continued growth of the world economy has depended on China's utilization of its massive coal reserves (its coal production is now three times that of the USA [3]). If the Western economies recover, oil shortages and rising energy prices are sure to follow.

The relationship between EROI and net energy is very much like that between interest rates and the price of a fixed rate bond. At high interest rates a 1 % move, let us say from 20 to 19 %, has a relatively small impact on the bond price. At low interest rates the effect is much greater; a move from 4 to 3 % will have a very significant impact on the bond price. As EROI falls below 10, the fall in net energy becomes much more significant; at 10 : 1 the net energy is 90 %, at 5 : 1 its 80 %, at 3 : 1 its 66 %. This has been referred to as the 'Net Energy Cliff'. Our modern societies have become so hooked on nearly-free energy that even a decline in the net energy below 90 % will cause severe issues, with an EROI of at least 8 : 1 being required to maintain the high living standards and complex society to which we have become accustomed (Fig. 1.2).

As more and more of the energy produced is required to provide new supplies, less and less is available for everything else, even if overall energy production remains stable. We are now close to the point where the economy will be directly

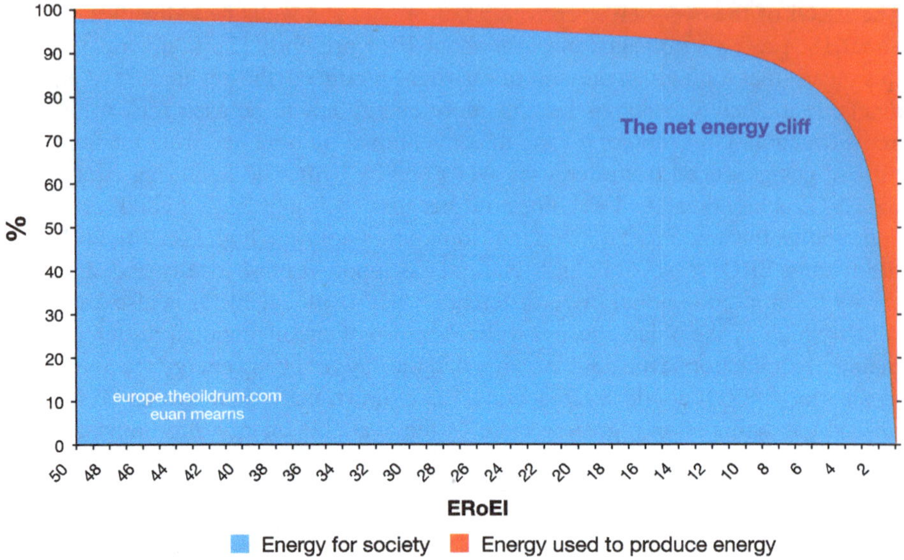

Fig. 1.2 Relationship between EROI and Net Energy. (Source: "The Net Energy Cliff" chart first developed by Dr Euan Mearns based on concept from Dr Nate Hagens.)

impacted, if in fact it has not already been so. Charles Hall and colleagues have produced a model of the impacts (below) which shows the difference between 1970 and his projections for 2030. Notice how a greater production of energy still results in a smaller level of consumption, as more of that energy has to be redirected towards its extraction and production. The biggest hit will be to discretionary consumption, with shrinking consumer and consumer finance sectors—no more Gucci in the shopping mall or bargains at Wal-Mart; perhaps no Wal-Mart or shopping mall at all. Also affected will be public investments, including the maintenance of current infrastructure such as roads and bridges. According to the Association of Civil Engineers, $ 3 trillion of investment is needed to return US bridges and roads to safe functioning levels [4].

Unfortunately, the newer oil and gas sources tend to have much lower rates of production, as well as lower EROIs. So, we get fewer barrels of oil while more of each barrel is used up in the extraction process. A telling comparison is that of the world's biggest oil field, Ghawar, with the Alberta tar (it really is tar, not oil) sands. The former has produced 5 million barrels of oil a day for many decades, using relatively straightforward drilling techniques. It has taken decades to get the tar sands to a production level of 2 million barrels per day (bpd) and even optimists see it growing to only 3.5 million bpd by 2025. Meanwhile, the Ghawar field is being flooded with vast amounts of sea water to maintain its output. This can only go on for so long before the inevitable drop in production arrives.

Since 2005, global oil production has been on a bumpy plateau, with the newer smaller flow and lower EROI fields struggling to offset the depletion of the much

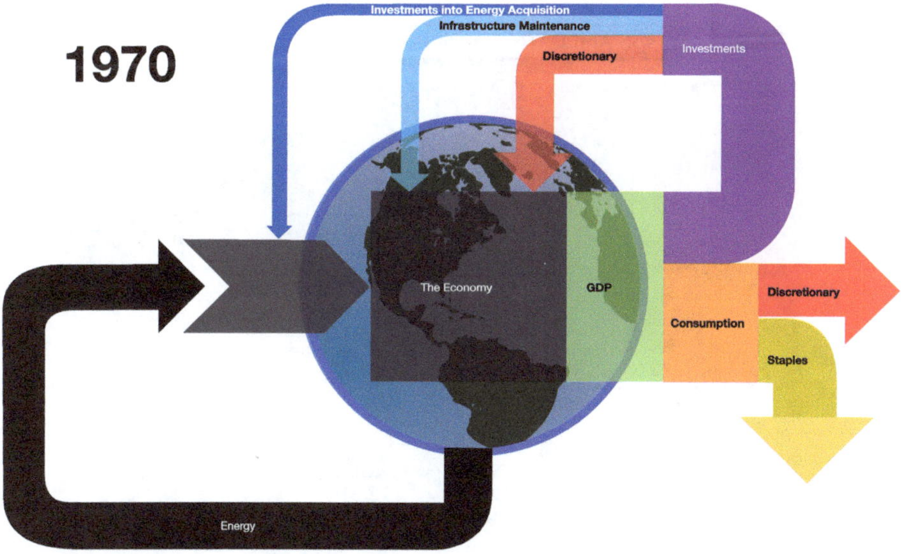

Fig. 1.3 Global Energy Usage and Production in 1970. (Source: Energy & the Wealth of Nations: Understanding the Biophysical Economy. Hall & Klitgaard, Springer 2011.)

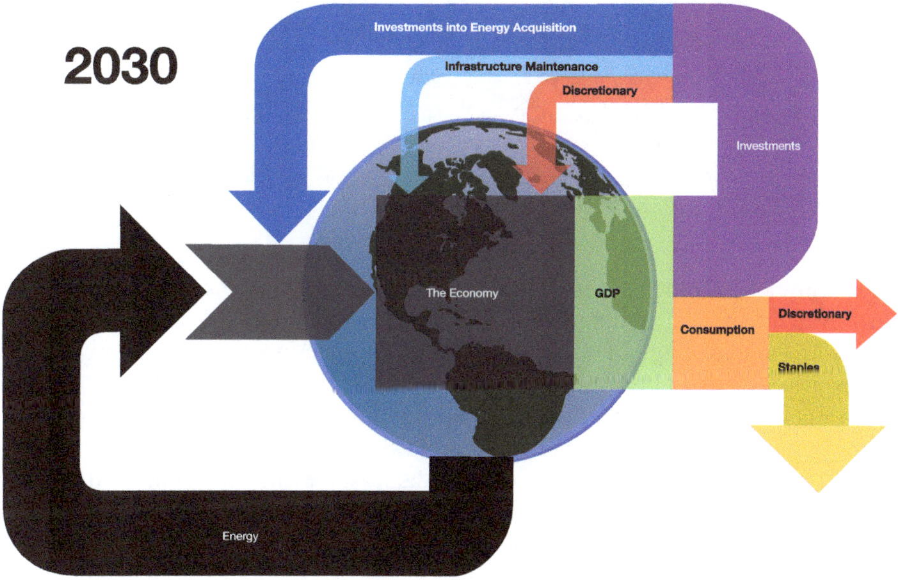

Fig. 1.4 Global Energy Usage and Production in 2030. (Source: Energy & the Wealth of Nations: Understanding the Biophysical Economy. Hall & Klitgaard, Springer 2011.)

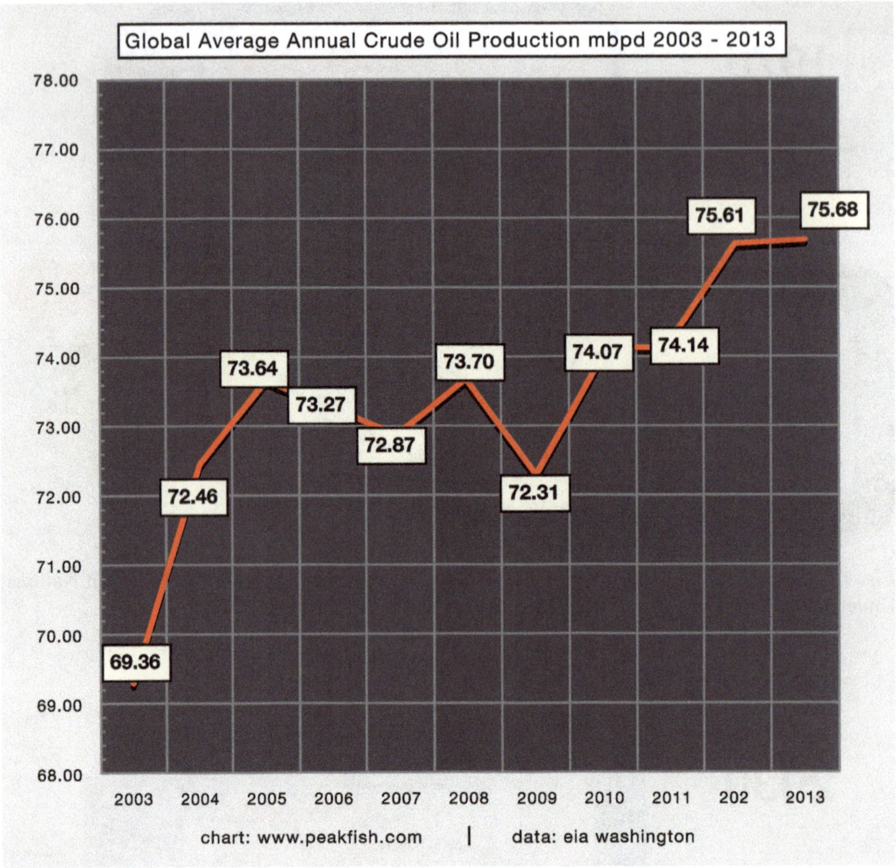

Fig. 1.5 Crude Oil Production 2002–2013. (Source: Global Average Annual Crude Oil Production mbpd 2002–2013. Peak Fish website article, September 25th, 2013)

larger flow of the older, higher EROI fields. Tar sands, fracking, deep sea oil, and now even attempts at finding Arctic oil are acts of increasing desperation to keep replacing the depleting older fields. If there were easier ways to get the oil, energy companies would not be bothering with these difficult operations. The new renewables, such as solar, PV, and wind are even poorer choices as they require much upfront investment and deliver their output in small incremental amounts over long periods of time, so for the first few years at the least they use more energy than they provide. When I put gas in my car, I get all its energy immediately, but solar panels and wind turbines deliver their energy output in small increments over decades (Fig. 1.5).

What all of this means for investors is that, at best, growth may cease at the global level in the relatively near future. Once you accept that growth will cease, all of the current 'common sense' assumptions about investing, such as the assumption of making money from money, cease to be true. Completely different assumptions will be required, including an understanding that the future will be a less wealthy place than the present. The modern financial system acts as a time machine, transferring our perceptions about the future into the present. Thus, rather than the actual arrival of non-growth, the simple failure of our general belief in continued growth, could destabilize and crash the financial system. Even if an end to economic growth were still 10 years in the future the impact upon financial assets could start being felt in the present as more and more people begin to accept its inevitability.

In this book, I first explain the concepts of Energy Return on Investment (EROI) and Energy Flow from the different energy sources in more detail. These two chapters are structured so that you can just read the introduction and the summary if you do not need to delve into the details of each energy source. Then I move on to explain the impact that growth constraints will have on financial wealth and the financial system. The final chapter covers what this means for individuals and their personal wealth. Whether you are a financial professional, or are simply concerned about your own wealth, this book will change the way you see the future, and hopefully allow you to keep some of the wealth you currently think that you have.

References

1. Grossman R (2011) Unsettled account: the evolution of banking in the industrialized world since 1800. Princeton University Press, Princeton
2. Hall C, Klitgaard K (2012) Energy and wealth of nations: understanding the biophysical economy. Springer, New York
3. World Coal Association (2012) Coal Facts. http://www.worldcoal.org/resources/coal-statistics/. Accessed 25 May 2013
4. Association of Civil Engineers (2013) 2013 Report Card for America's Infrastructure. http://www.infrastructurereportcard.org/. Accessed 25 May 2013

Chapter 2
It Takes Energy to Get Energy

2.1 A Very Short History of Humanity and Energy Return on Investment

Imagine you are a prehistoric hunter-gatherer. Your primary motivation is to find food, as that is your only source of the energy which your body needs to stay alive and without it you will slowly starve and die. The balance between the amount of food energy gained in hunting and gathering, and the amount of energy expended in those activities is critical. If that ratio is not high enough you will still die from starvation, just more slowly than if you had no food at all. Let us say that for every usable calorie of energy you gain, you expend 1 calorie in the hunting and gathering process. Then you would be fine, would you not? Unfortunately not, as you would still starve to death, just even more slowly. Your body has a whole host of functions which are not directly related to finding food but critical to your survival. First of all, there is the need for water, without which you will die in about 5–10 days. Then, you have to digest the food, to transform it into useful calories, and you have to excrete the processed food and water otherwise your bladder and intestines would swell up rendering you painfully immobile. So, let us say all these other functions require a doubling of calorific intake, so you now get 2 calories for every calorie spent hunting and gathering. Now you are good, right? Still the answer is no as the human body also has the annoying need to sleep for approximately one third of the day otherwise you will slowly malfunction and start going somewhat mad. So, let us add another calorie for that. Therefore, 3 calories gained for every calorie spent in hunting and gathering make for a good balance. Things are now good, right?

Actually, no because apart from you leading an extremely boring life, where every waking hour is spent searching for food and water, your branch of the human race would go extinct. Sex helps with both the boredom and extinction issues. Problem is that even if you are not into multi-hour Tantric sex sessions there is still some reduction in hunting and gathering time and there are those troublesome children to feed and spend yet more none food finding time raising. Let us say we now have two adults, which does tend to make sex a lot more fun, plus children who are only marginally useful at gathering food until their teens. Probably, we are in the 5:1 ratio of calories gained from calories expended for long-term survival. Add to

R. Boyd, *Energy and the Financial System,* SpringerBriefs in Energy,
DOI 10.1007/978-3-319-04238-1_2, © Roger Boyd 2013

that a lot of time for socializing with other relatives and clan members, and you get the kind of life style that our ancient hunter-gatherer ancestors enjoyed. The energy return on investment (EROI) for these hunter gatherers has been estimated at about 10:1, which allowed for a good amount of leisure time [1], needing only a few hours per day to find all the food they required. This did, however, involve a transient lifestyle, moving from place to place to take advantage of seasonal food sources, or after exploiting the food resources at a given place.

The first carbon-based fuels used by humans were predominantly wood and shrubs used to fuel a fire, with our ancestors learning to use and control such fire at least half a million years ago. A campfire could be used to ward off predatory animals, keep humans warm in colder climates, and to cook food which made it easier to digest while destroying dangerous parasites and microbes. Heating could also be used to dry and preserve foods for later consumption.

The hunter-gatherers that we would recognize as modern human beings, *Homo sapiens sapiens* (not a typo, this means, anatomically, modern humans), came along only about 200,000 years ago. It was not until about 50,000 years ago that some of the more advanced human traits developed, such as symbolic thought and standardized stone tools. For about another 40,000 years they continued to live in small groups of hunter-gatherers, and spread across the globe.

Approximately 11,000 years ago, some human groups took up farming. The experts disagree on whether this was forced upon them by a combination of environmental change and population growth, or it was a natural next step given our species' growing cleverness and tool-making capacities. Many hunter-gatherer groups do practice some control of the plants around them, for example by rooting out unwanted plants and clearing areas with fire to aid the growth of plants more beneficial to them. Thus, the move to agriculture may have been more of an incremental, rather than a revolutionary, change. What is not open to question is that the effects of this change were revolutionary, forming the basis upon which complex human societies have developed. Traditional agriculture has been estimated to have an energy return on investment [2] of about 15:1. Further, the ability to grow all the food required in a single place allowed for settled communities and, sometimes, a food surplus. Animal husbandry also developed, providing easily managed herds that could be slaughtered for consumption, dogs for hunting and to help manage those herds, and draft animals to provide the energy needed to pull ploughs. The surplus produced through agriculture and herding constituted excess energy beyond immediate needs which could be a source of power for those who controlled it. Agricultural societies started to diverge from the highly egalitarian hunter-gatherer groups into stratified societies where a majority of the population produced the food—and the energy surplus—while a minority controlled the use of that surplus. Those minorities became elites who could use the surpluses to feed individuals with new specialist skills such as metal workers, bureaucrats, and priests. They also fed fighting men to both control their tribe by force, where bureaucracy and religion failed, or to take over other tribes thus expanding the surplus available to them.

Over time these factors lead to the highly complex societies that started to appear about 6,000 years ago. It is within this incredibly small period of time, relative to the history of even our own species alone, what we refer to as *history* has taken place. The Egyptian, Mayan, Greek, and Roman civilizations, the Middle Ages, Renaissance, and the Industrial Revolution all took place in this brief period called *history*. Joseph Tainter refers to such complex societies as an "anomaly in history," as apart from this 6,000-year period. "Throughout the several million years that recognizable humans are known to have lived, the common political unit was the small, autonomous community [3]." However complex, these societies still relied upon the amount of food and other energy that could be provided by their surrounding ecosystem, and thus, were at the mercy of local climate changes and overuse of land. A further danger was posed by other societies which often went to war to gain new territory and slaves. Within these societies the tension between the poor masses that produced the surpluses utilized by the elites, and those elites, could also threaten to pull a society apart. Eventually the sheer complexity of these societies threatened their existence as ever greater amounts of energy were required to support them. In short, many factors could combine to greatly weaken a society; a few bad harvests could weaken military strength or exacerbate internal tensions as farmers struggled between feeding themselves and meeting the demands of the rulers. The fragility of such complex social arrangements has been demonstrated by their repeated collapses. Collapse awaited not only the relatively small and ecologically marginal societies, but also the highly developed ones at the peak of their power such as the Mayan, Sumerian, Greek, Roman and Khmer [4]. Civilizations could also be fractured into smaller parts: about 4,000 years ago during a 150-year drought that greatly reduced the Nile floods, Egypt was stretched to the breaking point [5]. The drought also seems to have played an important part in the collapse of the Bronze Age in the Eastern Mediterranean about 3,200 years ago, with complex societies such as the Mycenaean and Hittite collapsing into a "dark age" with much lower levels of political centralization and complexity [6, 7]. With a diversity of civilizations scattered thinly across the globe the failure of any individual civilization could not endanger human civilization as a whole. As civilizations rose and fell, humanity continued to grow in numbers, from 5 million people in 5,000 B.C., to 27 million in 2,000 B.C., doubling to 50 million by 1,000 B.C., and quadrupling to 200 million by 0 A.D.

About 5,300 years ago, in the Near East, Eastern Mediterranean, India, and China the need for energy greatly increased. The production of copper alloys, such as bronze required an intensity of heat that could only be provided by burning vast amounts of timber. This energy demand was further exacerbated by iron production, which started about 3,200 years ago, and required even higher temperatures that could only be provided by charcoal (wood heated in the absence of oxygen to produce a purer form of carbon). The resulting demands for timber as an energy source and for construction, plus the clearing of forest for agriculture, lead to the deforestation of huge areas as with the Mediterranean during the time of the Greeks and Romans.

The development of any civilization is limited by the amount of usable energy available to it, and the efficiency of energy use. The more energy and greater efficiency, the greater the percentage of the population that can be devoted to things other than sourcing that energy. More energy can also feed a greater the number of machines and extend human capabilities. Agricultural societies are limited by both the available biomass (predominantly food, fodder for animals, and wood) and the efficiency of the humans, horses, other draft animals in converting that biomass into useful energy. Human technology, through such things as the plough, wind-mill, and sail boat, can stretch that limit, but not fundamentally remove it. Defeating other societies, taking their land, and enslaving their people can also increase the amount of energy available. In most societies, small elites can live very well as they are able to capture the surpluses of the rest of the population. The celebrated democratic institutions of ancient Greece were only for a very small male minority, while most of the rest of the male population served as slaves and serfs. None of these factors removed the overall limit of complexity and development that agriculturally-based societies operate within. Ian Morris [8] developed an index of social development and noted that only three civilizations could be identified as reaching the low 40s on his index—the Song Dynasty, the Roman Empire, and modern civilization. As Morris [8] puts it "If someone from Rome or Song China had been transplanted to eighteenth-century London or Beijing, he or she would certainly have had many surprises… Yet more, in fact much more, would have seemed familiar… Most important of all, though, the visitors from the past would have noticed that although social development was moving higher than ever, the *ways* people were pushing it up hardly differed from how Romans and Song Chinese had pushed it up."

Only when human societies started to utilize the fossil fuels, with their high levels of energy density, was the limit upon their size and complexity removed. The Renaissance and the resulting scientific revolution may have stretched that limit, but without the utilization of fossil fuels they could not have broken free. With fossil fuels, the change in the energy available was revolutionary, driving the rapid changes to human civilization over the past two centuries. The ratio of energy gained to energy spent for the fossil fuels was at least 80:1 for coal, 100:1 for oil, and 18:1 for natural gas [9]. Also, the sheer volume of the energy that could be utilized dwarfed that which was previously available. Prior to this, the rate of energy use was limited by the depth and fertility of the soil, together with the vagaries of the weather. Now humanity had access to many millennia worth of photosynthesis which had been transformed into energy-dense substances and stored away under the ground. The only limitation was how quickly these new energy sources could be extracted. Naturally, the easiest ones went first. This huge increase in available energy has been the basis of modern industrial societies, which in turn have become addicted to this seemingly endless supply of cheap energy.

In this new age of fossil fuel, man has been referred to as the Anthropocene (anthropo means man) to delineate when humanity started to make a significant impact upon the ecology of the earth. Coal had been used in many areas including heating, after human demands lead to the destruction of forests, but the real revolution started when we worked out how to use coal to boil water into steam and to use this steam to drive an engine. These steam engines were first used in the coal mines to pump out water. In a few generations they found many others uses such as driving the wheels of trains and the propellers of ships, and driving turbines to create electricity. Coal could also be gasified to create town gas used for cooking and lighting. The available coal seemed endless, but the exponential growth facilitated by coal's concentrated energy drove an exponential demand for it. In the birthplace of the industrial revolution, the UK, the peak of coal production was reached in 1913 [9], just over a century after widespread adoption of the steam engine. Oil provided the additional energy needed for continued growth. It had been used as a heating and lighting fuel for thousands of years, but its combination with the internal combustion engine in the 1800s created the next revolutionary change. The use of the internal combustion engine expanded rapidly, producing a rapidly expanding demand for oil. As with coal, oil was seen as a nearly infinite resource, but the exponential growth in demand quickly reduced the huge amounts available. It took just over a century for the USA to reach its peak oil production, in 1972. The last part of the fossil fuel trinity, natural gas, had to wait until a transportation infrastructure of pipes was in place, which did not happen until the post World War II years. Since then its use has multiplied in the heating of homes, and the production of electricity. Humankind also discovered how to utilize these hydrocarbons as a raw material source in the production of new products, such as fertilizers, plastics, and organic chemicals which have become a ubiquitous and essential part of modern society. No matter how much we consider human ingenuity and the resulting technology to be the basis of our current living standards, the true determinant has been the finite fossil fuel resources that we first started to utilize extensively only a couple of centuries ago. Since that time our fossil fuels consumption has expanded nearly 800 times, with a 12-time increase during the twentieth century [10]. Without these fuels our modern society would not exist in its present form, and as we deplete them, we bring ever closer the day when we will have to do without them.

Some have argued that the advanced industrial countries have reduced their dependency on fossil fuels as more of their economies became service and computer based. This seems to be more a function of the advanced countries moving much of their energy-intensive manufacturing to countries such as Mexico and China. The energy intensity of what some countries produce may have gone down, but the energy intensity of their consumption which includes imported goods, has not. As for computers, a quick check of the data centers running an average Internet Service Provider, or Amazon, Facebook, or a major bank, will rapidly show the huge amounts of electricity that such computer-based organizations require, with much of that electricity provided by coal fired power stations—it is ironic, but the Internet ultimately dependent upon coal. Look around your own home at all those desktops, laptops, notebooks, tablets, cell phones, televisions, and gaming machines that need electricity. The growth of the economy and the complexity of our society are still very tightly linked to the energy available. Greater complexity tends to require greater levels of EROI, and lower levels of EROI lead to lower levels of complexity—i.e. simpler civilizations and economies. Less net energy equals poorer and simpler. As more energy is used to find and produce energy, Net Energy falls, meaning less is available to provide for anything else. And remember, 87 % of that energy is provided by fossil fuels [11].

2.2 The Fossil Fuels: 86.6 % of Human Society's Energy Usage

2.2.1 Oil: EROI 18:1 and Rapidly Declining; 32.6 % of Energy Usage (Excluding Ethanol and Biodiesel)

Oil provides 32.6% of global energy usage [11]. Oil's EROI had declined to an average of 18:1 globally by 2005 [10]. Oil is an incredible substance with a phenomenally high energy density, combined with an ease of transportation and storage given its liquid state at room temperature. This makes it perfect as a liquid transportation fuel, its primary use in society today. A tank of gas costing C$ 65 was all that was needed to propel my family's Mini Cooper along the 300 miles (about 500 km) from Toronto to Montreal at speeds of up to 80 miles/h (about 130 km/h), with some gas to spare. Such things would have seemed miraculous only a few generations ago, but we now take them for granted. The cost of this wondrous substance has recently increased substantially. That same tank of gas cost only C$ 30 about a decade ago, and prices will rise further as oil from the newer sources costs much more than from the earlier ones.

Just like with an apple tree, where the lowest hanging apples tend to be taken first, the simplest and cheapest ways to produce oil were used first while the more complex and expensive ones were only exploited when needed. At the beginning, oil drilling only required a hollow steel tube to be pushed into the ground to a relatively shallow depth and the oil would gush out under its own pressure. Now that scene is confined to old movies where oil is struck and gushes up into the air with everyone jumping up and down for joy as they are covered in oil. Later, after the pressure normalized, a "nodding donkey" would be used to suck up the oil. The world's oil supply is still very dependent upon many of those older, easy-to-extract oil fields. Nearly half of the world's oil supply is provided by just over 120 giant oil fields, the majority of which were discovered many decades ago. In fact, nearly 70% of current global oil production comes from fields discovered before 1970. These less expensive, high net energy fields, keep the average global EROI at a high enough level to support our complex civilization. As these fields age and their EROI falls, technological fixes such as horizontal drilling and pumping in water to drive the oil to the surface are needed. Their depletion subtracts relatively high net energy oil from the overall supply. The greatest of these mature fields is Ghawar in Saudi Arabia which has been operating since 1951 and still provides 5 million barrels of oil per day, or about 6% of global oil production. Ghawar's output is about three times the 2012 production of the Canadian Oil Sands, which is optimistically forecast to produce 5 million barrels per day—but not until 2030. Unlike Ghawar, some of the other large conventional oil fields have seen very significant drops in production as they deplete, such as Cantarell in Mexico, Prudhoe Bay in the USA, and Samotlor in Russia. Recent oil finds are much less accessible and much more expensive to produce. Instead of gushers and simple "nodding donkeys" we have the complexities and costs associated with such things as ultra deep sea oil fields, tar sands, and shale oil.

The recently discovered large Lula ultra deep sea oil field is 160 miles off the Brazilian coast, below approximately 6,600 ft of ocean and 16,000 ft of salt, sand, and rock. Compare that to the world's tallest building which is only 2,700 ft in height. At these depths, the pressure within the piping rises as high as 15,000–20,000 pounds per square inch, not far away from the pressures used in industrial high pressure water machines used to cut metal. The drilling of the first 15 wells for Lula cost over US$ 1 billion, and the total field costs are estimated to be in the US$ 50–100 billion range. With development covering an area the size of Florida, the Canadian tar sands make this operation look small scale—and technically simple. The Canadian tar has to be first separated from the sand then mixed with other hydrocarbons to make synthetic heavy oil. When surface mined, it takes 2 t of tar sand to produce 1 barrel of synthetic crude oil; hence the army of huge trucks. The separation process also requires about 2–4 barrels of hot water and caustic soda per barrel of synthetic oil. With the cheaper surface deposits already being heavily exploited, the focus has moved to the deeper deposits which require drilling. To turn the tar sands into a liquid which will travel along the drill pipes, huge amounts of heat and hydrocarbon solvents are required. The heat required has been predominantly provided by the burning of natural gas, using up one fossil fuel to gain access to another. Another recent phenomenon is the production of shale oil using hydraulic fracturing (or "fracking") technology. This involves drilling a hole into a rock formation and then injecting large amounts of a complex mixture under pressure to fracture the rocks and thus free oil trapped within them. Each drill site produces relatively small amounts of oil and depletes much faster than conventional oil wells. This means drilling hundreds, then thousands of expensive new wells to maintain production. On average, the ratio of energy gained versus the energy expended for ultra deep sea, tar sands, and shale oil is about 5:1 or below—bad news for the future of our complex society.

There is also oil shale (which is different from shale oil). Oil shale does not contain oil, but rather Kerogen, a chemical precursor to oil. The EROI of mining this shale and turning it into synthetic oil has been estimated at 2:1 [12]. The burning of the oil shale in an electric power station has been proposed, but the low energy density of the oil shale would produce massive amounts of waste products. Synthetic oil products can also be produced from coal, a complex process called coal to liquids (CTL), but again, the conversion from coal to oil is a very energy intensive process reducing the EROI to about 4:1 [13]. CTL has only been utilized by countries that could not access enough conventional oil supplies, countries such as Germany in World War II and South Africa during the anti-apartheid oil embargo. The gas to liquids (GTL) process, where synthetic oil products are derived from natural gas is hardly any better. The use of these technologies is either a sign of desperation, or of large price discrepancies between oil and natural gas caused by a localized oversupply of natural gas, as is currently seen in North America.

Oil provides about a third of all the energy consumed in the world. It powers the vast majority of all the trains, planes, cars, trucks, ships and motorcycles in the world, and it provides chemical inputs to a large share of important industrial and consumer products. Its loss to humanity could render modern society non-viable. For the moment—and it is just a moment in human history—humanity is living off the remains of the easily accessible oil fields which produce very high Net Energy returns. But as those fields become depleted the average Net Energy for oil falls. In 2005 the EROI for global oil production was 18 to 1, and has quite possibly already fallen below the 10:1 danger point given the historical declining trend. That is an average of all fields, while the net energy available from new fields may be approaching the point of being economically unviable. These new fields require increasing levels of complexity and investment levels with many unknowns. Possible delays and other unforeseen problems could easily invalidate the cost assumptions that once made them appear profitable. A vivid example is that of the Deep Water Horizon. What will be that well's net energy level after all of the rescue and remediation costs are taken into account? The desperation of oil producers to find new oil to replace the depleting older wells has been shown by the attempts to drill in the Arctic by Shell Oil, even before all the ice has gone. Drilling was repeatedly impacted by drifting ice, and then a near hurricane in the North Pacific set the oil rig adrift for a few days [14]. Yet, at last report, Shell has vowed to return.

2.2.2 Coal: EROI 20-80:1 and Slowly Declining; 30.3 % of Energy Usage

Coal provides 30.3 % of global energy usage [11], and has an EROI ranging from 40 to 80 for the USA and about 20 for China, depending upon the mining conditions and the quality of the coal [10]. It is less energy dense than oil and is a solid which makes transport much harder and more expensive (trains and ships instead of pipelines). Coal also burns less cleanly than oil or gas. These factors limit both its use and the distances that it can be economically transported. Unlike oil, most coal is used within the region in which it is mined, with limited exports between countries. Even in China, the biggest importer of coal, imports represent less than 10 % of its consumption with a significant amount of those imports coming from neighbouring countries. Coal's use as a transportation fuel for ground and sea transportation was superseded by the much more efficient oil in the early twentieth century. Presently its main use is in power stations to produce electricity, and secondarily in the production of steel. The USA uses coal to produce about 40 % of its electricity.

By far the biggest user of coal is China, which gets approximately 80 % of its energy from coal and uses more than three times as much coal as the USA [15]. The sheer speed at which Chinese coal consumption has grown is shown by the fact that its coal use was only twice that of the USA just 5 years ago [11]. It has added the amount consumed by the whole of the USA in only about 5 years! Coal is not a homogeneous substance; rather, it is grouped into four general categories which correspond roughly to energy density. These are anthracite (the best), through bituminous and sub-bituminous, to lignite (the worst). At a lower energy density, more coal has to be mined to provide the same energy output. In addition, coal is found at different depths, and with different seam widths. As the depths increase and the seam widths decrease more energy is required to mine the coal, again reducing its EROI. As with oil, the easiest coal deposits were mined first, and thus, the quality and accessibility of coal deteriorates over time. As lower quality and less accessible ores become a greater and greater percentage of production, the amount of coal produced may not be a good indication of the energy being provided. This has been the case with the USA where increases in coal production have paralleled a *decreased* amount of energy provided due to ongoing reductions in the energy density of the coal.

With oil production at a plateau since 2005, and the Chinese focus on their huge coal reserves to fuel growth, coal is poised to regain the spot of the leading world energy source that it lost to oil in the twentieth century. This trend can only continue with the huge India population looking to fuel industrialization with their own coal reserves. In 2011 global energy use from coal grew by 5.4 %, but only 0.6 % for oil and 2.3 % for natural gas [8]. In 2012 Chinese coal imports increased 20 % to 290 million tons [16] although they aim to limit their coal consumption to "only" 3.9 billion tons by 2015 [17] an amount nearly 4 times that consumed by the USA. This is twice bad news for the environment not only because burning coal releases a lot more climate warming carbon dioxide and other poisonous gases than oil, but also because the lower the quality of coal, the more that is burnt for a given amount of energy. Opponents may have stopped the construction of new coal fired power stations in the USA, but their gains are overwhelmed by new coal fired plants in China and India. Carbon Capture and Storage (CCS), where the carbon dioxide produced by burning the coal is captured and pumped into huge storage areas such as depleted oil fields or the deep ocean, has been proposed by the coal industry to remedy the climate changing impacts. If utilized, this process is very energy intensive requiring 25–50 % more coal to produce the same energy output, thus reducing the energy available to society. So far, there are very few actual working examples of CCS to prove its viability.

2.2.3 Natural Gas: EROI 10:1 and Declining; 23.7 % of Energy Usage

Natural Gas supplies 23.7 % of global energy usage [11], and has a current global EROI of about 10:1 for conventional gas fields [10]. It was first treated as a useless by-product of oil extraction, being "flared off" at the well site. No one has documented how much natural gas was wasted in this way, but the amounts must have been huge. The use of natural gas has grown rapidly over the past few decades. It is increasingly being used for space heating, cooking, generating electricity, and in the production of synthetic fertilizers. The latter use played a large part in the post-war "Green Revolution" which facilitated the rapid growth of populations in many countries. Being a gas at room temperature NG is primarily moved overland via pipelines. To transport NG by sea it has to be cooled to −162° C at which point it becomes a liquid that can be carried in specially engineered liquid natural gas (LNG) ships. This is an expensive process which offsets some of the energy provided by the natural gas thus transported. These comparative factors support the construction of pipelines over many thousands of miles between Russian production fields and Europe via the Ukraine and Poland. Despite these limitations, LNG has grown rapidly to become a quarter of the global natural gas export market [4].

Presently many countries with large NG supplies cannot utilize it all, a number of richer countries have heavily depleted their fields, and there are large pricing discrepancies between regional markets separated by water. It is currently two to three times as expensive to buy natural gas in Europe as it is in North America while prices in the Far East are even higher. One possibility in "gas rich" countries is to use LNG as a transport fuel to replace oil, a strategy which has been successfully used in buses, trucks, and ships. This could definitely be an option to reduce the impact of declining cheap oil reserves, providing some more time for society to find alternative fuels and living arrangements. On the other hand, if such substitution was applied extensively it would significantly increase the demand for natural gas and thus speed up depletion of positive net energy reserves.

Another source of natural gas, shale gas, requires the use of fracking techniques (see oil fracking above) to exploit gas reserves that were otherwise not viable. This has greatly changed the dynamics of the market, especially in North America where conventional output was in significant decline. This new unconventional supply has caused a glut in North America due to government restrictions on exports. The resulting substantial fall in North American natural gas prices has lead to its greater use in the production of electricity, displacing coal. However, this may not be a tenable situation in the medium term as many current shale gas wells are unprofitable at anywhere near the current price. Shale gas wells also have very high rates of production decline over time, much higher than for conventional gas; greatly reducing the likelihood of production paying back the US$ 10 million that such wells cost. One researcher has estimated that the break even cost for shale gas is around US$ 9 per million British Thermal Units, rather than the current US$ 3.50 [18]. Further complicating the situation may be supply constraints as the profitable finds may be limited to relatively small sweet spots within an overall gas field. For shale gas estimates for EROI vary greatly, from as low as 5:1 to as high as 70:1 and above. This may reflect differences in how EROI was calculated for what is a relatively new resource, and the possibility of "sweet spots" where well productivity is much higher. On the plus side, natural gas does have a reputation for being more climate friendly than coal, with lower amounts of carbon dioxide being released when burnt. However, studies showing that significant amounts of methane can leak from the drilling and transportation processes [19] have undercut this advantage. While methane only stays in the atmosphere for about 14 years, for each of those years it can have a climate changing impact of up to one hundred times that of carbon dioxide.

2.3 Hydro and Nuclear: 11.3 % of Global Energy Usage

With the huge increases in available energy that these three fossil fuels have provided it has also been possible to develop other ways of gaining access to useful energy. The major ones have been the damming of rivers and the exploitation of nuclear energy.

2.3.1 Hydroelectricity: EROI 12:1 to 267:1 and Stable; 6.4 % of Energy Usage

Hydroelectricity supplies 6.4 % of global energy usage [11]. The net energy of hydroelectric dams varies greatly due to the specifics of each site, and range from 12:1 to 267:1 [10], all providing enough net energy for an advanced civilization. The vast majority of the energy expenditure on hydroelectricity is during the construction phase so existing dams have very high net energy levels. Most of the good dam sites in North America have already been utilized, but there are many unused sites around the world. As large areas are flooded, big dams create major social impacts which may limit the use of many possible sites. China has leapt heavily into dam building despite frequent widespread displacement of local populations, as with its huge Three Gorges dam project.

Hydroelectricity can also be provided through "run of the river" projects, where the ongoing river flow is used to drive turbines. These can be relatively small installations and are good candidates for providing localized electricity supplies. With such local supplies, complex electricity grid infrastructures are not needed to transfer the electricity from producer to consumer. In addition, the natural flow of the river is kept, fish populations are not impacted, and valuable land is not flooded.

Winter snow acts as a huge natural water storage device, many times the volume of any dam, keeping vast water supplies in place until the spring and summer melts. Glaciers provide the same service, increasing their flows during the warmer months. This is especially important in the sub-tropics, where the resulting river flows make up for a lack of local rainfall during the summer months. As climate change induced warming changes winter snows into rain, further shrinks the glaciers, and alters the "spring thaw" calendar, these storage mechanisms will become unreliable. The result will be much less dependable river flows available to hydroelectric facilities, possibly reducing the amount of electricity that can be produced economically. Warmer temperatures will also produce greater levels of evaporation from dam reservoirs which will reduce the water available to drive turbines and produce electricity. Electricity production could even cease entirely if the water level drops below the water intakes for the turbines. Many dams in the American Southwest, such as the Hoover Dam, and those on the heavily dammed rivers that flow from the Himalayas through China, India, and Pakistan will be affected. Brazil has already shown the possible impacts, as an ongoing drought reduced hydroelectric output at the start of 2013 [20]. In other areas, climate change may provide for more precipitation (both rain and snow) and thus increase the productivity of some hydroelectric dam sites. Probable beneficiaries in this regard are Canada, Northern Europe and Northern Russia.

2.3.2 Nuclear: EROI About 5:1 with Large Uncertainties; 4.9 % of Energy Usage

Nuclear power currently provides 4.9 % of global energy usage [11]. Of all energy sources, nuclear power is easily the most controversial, with the Chernobyl and Fukushima crises showing the risks of using nuclear fission to produce electricity. Among the casualties of such large and costly failures has been commercial insurance which is no longer a viable option for the nuclear industry. This hurdle has been overcome by governments taking on the responsibility for the excessive costs of accidents, or building and operating the plants themselves. In addition to insurance, no nuclear power station has ever been built without significant government backing, let alone all of the government funded research provided free to the industry. And when accidents happen many of the costs tend to be socialized as they overwhelm the ability of any private company to meet them. Other social costs include the victims who are also not compensated since cancers happening years later cannot be definitively linked to a single given cause. In truth, nuclear power is not competitive in a free market situation.

These large subsidies, as well as the cost of storing depleted nuclear material possibly for thousands of years into the future, are not reflected in industry cost structures. Given these very hard to account for cost issues, plus the highly divisive and politicized nature of nuclear power, it is very hard to gain an accurate EROI for nuclear power. Charles Hall [10] estimates it at probably no greater than 5:1, which is well below that needed to support our current civilization's level of complexity. Recently much of the fuel for American reactors has been provided through the conversion of American and Russian nuclear warheads. With this source of supply dwindling, there will be a greater need for newly mined uranium—a depleting resource with costs rising as lower grade ores are utilized. Fast breeder reactors promise more efficient use of the source materials, but also involve even greater levels of engineering complexity than conventional nuclear reactors, and no commercially viable fast breeder reactor has ever been built. There have been other possibilities proposed, such as Thorium-based reactors and the TerraPower traveling wave design, but these are still either on the drawing board or at the early prototype stage at best. The time from where they currently stand to large scale usage is possibly counted in decades.

The impact of the Fukushima crisis has been significant, with Germany committing to close down all nuclear generation and Japan doing the same. Global production of electricity from nuclear fuel actually fell 4.3 % in 2011 with the sharp declines in those two countries [11]. In addition, the Canadian province of Quebec recently decided to retire its only nuclear electricity plant and the state of Vermont reached a deal to retire the aging Vermont Yankee plant. Both China and India have major programs of nuclear power station development, but at the global level any new production will be significantly offset by the probable retirements of existing stations which are on average 26 years old [21]. The USA also has plans for new nuclear power stations, but popular resistance and the long permitting process will significantly limit the rate of new development.

2.4 Non-Hydro Modern Renewable Energy: 2.1 % of Global Energy Usage

In recent years renewable sources of energy, such as wind, solar, biofuels and wave, have been developed. These renewable resources still only constitute a small share of global energy supplies however, increasing from 0.7 % in 2001 to 2.1 % a decade later [11]. Some of these are more promising than others as candidates to provide the net energy rates required for our society to continue in its present form.

2.4.1 Wind: EROI Averaging 18:1; Approximately 1 % of Energy Usage

Humans have used wind energy for many centuries with the first windmills being used at least 1000 years ago, if not much earlier. In recent decades, the primary focus of new wind energy implementation has been to produce electricity, thus reducing the need to burn coal and natural gas in electric power stations. Wind power currently provides approximately 1 % of global energy usage [11].

As with property, the success of a wind energy installation is all about location, location, location. A wind farm needs to be where the wind is strong and reliable enough to drive the wind turbines, and both of these factors vary greatly from one place to another. Construction costs, together with operating delivery costs, will also be greatly affected by the distance between the wind farm location, population and/or industrial centers, and the supplies of the required components and skilled technicians. These factors drive the profitable size of a wind farm or of individual turbines. Larger wind turbines or a greater number of turbines in one farm tend to reduce some of the fixed and ongoing maintenance costs relative to the electricity produced.

These factors tend to benefit offshore wind farms which are easier to locate closer to population centers and are able to utilize more and larger wind turbines—plus the wind tends to be stronger and more reliable offshore. All of these benefits usually outweigh the extra complexities of building wind farms offshore. The coasts of the Atlantic Ocean, the North American Great Lakes and the Baltic Sea tend to be relatively shallow, providing good locations. The profitable size and power output for both wind farms and turbines may be lower in more compact countries, such as those in Europe, as average distances tend to be much less than in larger countries such as those in North America.

Unfortunately, wind energy cannot provide a reliable "base" level of electricity as it is dependent upon wind which can vary in strength and direction. To balance out the fluctuations of wind energy another energy source is required to provide a reliable "base load". Some countries, such as Spain and Denmark, have managed to get up to 20 % of their electricity generated by wind energy while using other sources of electricity for the base load. Germany, which currently uses wind energy for about 10 % of its electricity is targeting up to 25 % by 2020. Presumably they will use a mix of conventional and alternative energy to provide the "base." With a diverse set of renewable energies used, and wind farms located in varied locations with differing wind speed peaks and troughs, it could be theoretically possible to do without hydrocarbon fuels for the base load. There are many technical challenges though, that would have to be worked out before such a setup would be truly feasible.

Wind energy installations are growing rapidly, and this growing demand is fuelling rapid growth, greater efficiency, and technical advances in the industry. In 2011, wind energy produced approximately 240 GW of electricity globally (about 3 % of global electricity production), up about 20 % from the previous year, having grown at a compound rate of over 20 % for many years. It is forecast to keep growing by 15–20 % globally ongoing, doubling by 2016 [22], and doubling again by 2020 to 1,000 GW. China is very rapidly driving installations of wind energy to help serve its growing power needs, and with 3 % of its electricity coming from wind energy it has advanced, in only a few years, to having 26 % of all global installed capacity in 2011. The USA is second with 20 %, Germany third with 12 %, followed by Spain with 9 % and India with 7 % [23].

2.4.2 Biofuels: EROI of 1:1 to 18:1; Approximately 0.5 % of Energy Usage

Biofuels, such as ethanol and biodiesel currently provide approximately 0.5 % of global energy usage [11]. The EROI [10] of corn ethanol is close to 1:1, and it would surely not be a viable fuel source without the large government subsidies and mandates provided to its suppliers, especially in the USA. EROI [24] estimates for sugar cane ethanol range up to 10:1 in Brazil. Unlike corn and sugar cane ethanol, cellulosic ethanol is not in large scale production, and thus there is both a wider variability for input and output assumptions, and for possible future technical and scaling efficiencies. There are very wide variations in the estimates for the EROI of cellulosic ethanol, from such plants as switch grass and willow. The EROI [24] estimates range from as low as 1:1, to as high as 18:1. That of biodiesel [10] is about 3:1.

The EROI of all of the biofuels is significantly affected by growing conditions, as the biomass yield is a significant determinant of the net energy provided. Given the great variability in growing conditions for different geographic locations, the net energy of biofuels will also greatly vary between locations. Yields may also be significantly reduced by the effects of Climate Change, with less water, generally higher temperatures, and less predictable weather patterns. Such issues have become apparent recently in the USA where ongoing drought conditions reduced plant yields and have lead to calls for a reduction in corn ethanol production or usage mandates. With biofuels competing for acreage with food production, there is increasing conflict between the drive to produce such renewable fuels and the need of the world's population to secure enough food. The losers are those with lower incomes as food prices escalate due to reduced supplies.

2.4.3 Solar: EROI Between 2:1 and 12:1; Less Than 0.5 % of Energy Usage

The energy output for solar technologies is location specific, with areas nearer the equator having longer days and stronger sunshine year round compared to more northerly and southerly locations. This means the most productive possible locations are in areas such as the American Southwest and Saharan Africa that are distant from the populations that would use the power. To exploit this energy and deliver it to consumers, extensive investments in new high-voltage long-distance electric power lines would be required.

An additional issue is that solar power installations are assumed to operate for at least a couple of decades to arrive at the computed net energy levels. Over such a period the operation of "Murphy's Law" has a great amount of time to impact the assumed output; through such things as sand storms, greater component failure rates, theft and other unpredictable factors. For example would low income communities happily coexist with solar installations and long distance power lines that were to serve far away populations rather than themselves? At the least, this could create significant security costs, as with those in many oil and gas producing areas of the Middle East and Africa. A large solar power installation would be just as vulnerable as Nigerian oil pipelines or the Algerian natural gas plant attacked in early 2013 [25].

Solar PV uses solar panels that convert the sun's energy into electrical energy. Estimates of energy return for Solar PV [26] range between 6:1 and 12:1, although some researchers have calculated an EROI as high [27] as 38:1, and as low [28] as 2.5:1. Differing assumptions such as the useful life of the solar PV installation, the maintenance costs over time, and the range of input costs have driven these wide variances. Solar PV can be used in large scale industrial implementations, as well as for individual commercial and residential buildings.

Concentrated Solar Power utilizes arrays of mirrors or lenses to focus solar energy onto a single small area, where a heat engine utilizes the thermal energy to create steam that drives an electrical generator. Solar thermal does not produce electricity, but instead uses the sun's heat to directly warm water or the air in living spaces. This can be used for commercial buildings and private dwellings. Both Concentrated Solar and Solar Thermal have an EROI [26] between 5:1 and 2:1.

2.4.4 *Wave: EROI Not Available; Negligible % of Energy Usage*

The wave power industry is in its infancy, with a very limited set of installations and much uncertainty about how actual installations will behave during long term exposure to corrosive sea water and the affects of storms.

2.4.5 *Geo-Thermal: EROI Not Available; Negligible % of Energy Usage*

Geo-thermal power has only limited installations worldwide and is highly location specific, being most effective in areas near tectonic plate boundaries. Iceland, with a population of about 350,000, is the only country which currently derives a majority of its energy from geo-thermal sources.

2.5 Summary

Large scale industrial societies came into existence through the exploitation of phenomenally large fossil fuel energy sources which were both energy dense and cheap to access. The net energy derived from fossil fuels was very high, with energy return on investment ratios of 80:1 and more. The massive scale of these new resources combined with this high net energy has provided the huge amounts of energy required to drive our modern societies. Without fossil fuels the industrial revolution and the related advances in living standards would likely have stalled at the wind and water power stage of development.

The easiest to access fossil fuel deposits tended to be the first ones to be used. As these depleted, and more and more energy was required by exponentially growing human societies, more difficult deposits which provide lower amounts of net energy began to be used. This has led to an ongoing reduction in the net energy provided by oil and natural gas deposits. The most worrying of these is the fall in net energy of oil production, as this fuel powers the vast majority of global transportation, and provides one third of all energy used by human societies. Its EROI was approximately 18:1 globally in the late 2000s and continues to fall as lower and lower net energy deposits such as tar sands, deep ocean, and shale oil are brought into production. The net energy for natural gas is also falling with the exploitation of shale gas and the costs of transporting it in liquid form between continents. Natural gas may be useful in the short term as a substitute for declining cheap oil reserves, but if such substitution was done on a significant scale, the increased demand would deplete the viable natural gas reserves at a much faster rate than otherwise. Only coal has continued to provide a relatively stable level of net energy, but its contribution to global warming argues against increased use.

A hierarchy of minimum levels of societal net energy has been proposed for the support of the more complex parts of industrial societies [26]; with a societal EROI of up to 14:1 required to support such things as education, health care, and the arts. Newer oil and natural gas production may already be below this level, with coal and the older oil and natural gas fields providing the required net energy to support current living standards. As the older oil and gas fields continue to deplete, the overall level of net energy available to society will fall below that needed to support current living standards. Coal will be left as the only fossil fuel that can provide the needed levels of net energy. Faced with unacceptable environmental degradation, societies will have to simplify with living standards changing to match the net energy available.

The alternative is to find new energy sources which provide the required levels of net energy to support an advanced industrialized society without undue environmental costs. Of the alternatives only hydro, wind, some biofuels, and possibly solar PV, provide the right level of net energy. Hydro is limited by the availability of dam sites, most of which in Europe and North America are already being used. Through the usage of dam sites across the developing nations a doubling of current capacity is possible to about 12 % of global energy usage, but has to be balanced against the impact of climate change upon the output of some sites. Although important, such an increase would make up for less than one tenth of the energy currently provided by fossil fuels. The remainder would need to be met by wind, selected biofuels and solar PV, which currently provide less than 2 % of global energy needs. Such a transition will require continued rapid growth in new installations, together with extensive changes to energy infrastructures and end-user consumption. Historically such changes have required decades, as with the change from coal to oil for transport which involved the replacement of a much smaller amount of energy than would be the case for a fossil to non-fossil energy conversion.

Another complicating issue is that hydro, wind, and solar PV installations require large up-front investments, which draw on current energy supplies, while the resulting energy input is delivered in increments over decades, taking many years to offset the up-front energy investment. Given that very rapid growth is assumed for such installations over the next decade and more, the new renewable sector may not provide a net positive provision of energy to society for many years. The large investments in renewables may, in fact, require the reduction of energy use in the rest of the economy during the extended "build-out" period. In effect, society may have to accept static if not lower living standards to allow for the redirection of energy to the construction of the new energy sources.

References

1. Heinberg R (2009) Searching for a miracle, Post Carbon Institute. http://www.postcarbon.org/new-site-files/Reports/Searching_for_a_Miracle_web10nov09.pdf
2. Fischer-Kowalski M et al (2011) Sociometabolic transitions in subsistence communities: Boserup revisited in four comparative case studies. Hum Ecol Rev 18(2):2011
3. Tainter J (1988) The collapse of complex societies. Cambridge University Press, Cambridge
4. Diamond J (2004) Collapse: how societies choose to fail or succeed. Viking Adult, New York
5. Fagan B (2003) The long summer. Basic Books, New York
6. Bachhuber C, Roberts G (2009) Forces of transformation: the end of the bronze age in the mediterranean (BANEA Monograph). Oxbow Books, Oxford
7. Dickinson O (2007) The Aegean from Bronze Age to Iron Age: continuity and change between the twelfth and eighth centuries BC. Routledge, London
8. Morris I (2010) Why the West rules—for now: the patterns of history, and what they reveal about the future. Farrar, Strauss, and Giroux, London
9. N/A (2011) Jevons' coal question: Why the UK coal peak wasn't as bad as expected. The oil drum. http://www.theoildrum.com/node/8241. Accessed 27 Dec 2012
10. Hall C (2012) Energy and the wealth of nations: understanding the biophysical economy. Springer, New York
11. N/A (2012) BP statistical review of world energy June 2012, British Petroleum. http://www.bp.com/assets/bp_internet/globalbp/globalbp_uk_english/reports_and_publications/statistical_energy_review_2011/STAGING/local_assets/pdf/statistical_review_of_world_energy_full_report_2012.pdf. Accessed 10 March 2013
12. Mearns E (2012) Unconventional oil and gas, a game changer? Presented at the 2012 Association for Peak Oil conference. http://www.aspo2012.at/wp-content/uploads/2012/06/Mearns_aspo2012.pdf. Accessed 12 March 2013
13. Talberth J (2009) Economic feasibility of coal to liquids development in Alaska's interior, Centre for Sustainable Economy. http://www.sustainable-economy.org/main/news/1. Accessed 10 March 2013
14. Efstathiou Jr J (2013) Rig grounding revives debate over Shell's arctic drilling, Bloomberg. http://www.bloomberg.com/news/2013-01-03/rig-grounding-revives-debate-over-shell-s-arctic-drilling.html. Accessed 15 March 2013
15. N/A (2012) Coal facts 2011, World Coal Association. http://www.worldcoal.org/resources/coal-statistics/. Accessed 25 May 2013
16. N/A (2012) Yearbook 2012: global market energy review, Enerdata. http://www.enerdata.net/enerdatauk/press-and-publication/publications/world-energy-statistics-supply-and-demand.php. Accessed 9 March 2013
17. N/A (2012) China to restrict coal demand, output to 3.9 billion tons, Bloomberg. http://www.bloomberg.com/news/2012-03-22/china-to-restrict-coal-demand-output-to-3-9-billion-tons.html. Accessed 10 March 2013
18. Melling A (2010) Natural gas pricing and its future: Europe as the battleground, Carnegie endowment. http://www.carnegieendowment.org/files/gas_pricing_europe.pdf. Accessed 11 March 2013
19. Romm J (2013) Bridge to nowhere? NOAA confirms high methane leakage rate of up to 9% from gas fields, cutting climate benefit, think progress. http://thinkprogress.org/climate/2013/01/02/1388021/bridge-to-nowhere-noaa-confirms-high-methane-leakage-rate-up-to-9-from-gas-fields-gutting-climate-benefit/. Accessed 20 May 2013
20. Boadle A (2013) Brazil's hot, dry summer may lead to energy rationing. Reuters. http://www.reuters.com/article/2013/01/07/us-brazil-energy-rationing-idUSBRE9060TL20130107?feedType=RSS&feedName=environmentNews&utm_source=feedburner&utm_medium=feed&utm_campaign=Feed%3A+reuters%2Fenvironment+%28News+%2F+US+%2F+Environment%29. Accessed 11 May 2013

21. Schneider M, Froggart A, Thomas S (2012) The World Nuclear Industry Status Report 2010–2011. Worldwatch. http://www.worldwatch.org/system/files/NuclearStatusReport2011_prel.pdf. Accessed 11 May 2013
22. N/A (2012) Global statistics. Global Wind Energy Council. http://www.gwec.net/global-figures/graphs/. Accessed 15 May 2013
23. N/A (2012) Cumulative installed wind power capacity in top ten countries and the world, 1980–2011. Earth Policy Institute. http://www.google.ca/url?sa=t&rct=j&q=&esrc=s&source=web&cd=5&sqi=2&ved=0CFEQFjAE&url=http%3A%2F%2Fwww.earth-policy.org%2Fdatacenter%2Fxls%2Findicator10_2012_all.xls&ei=BHLuUO6CDs3aqQHxuYC4BA&usg=AFQjCNE2fx8PmPAP2055qt-Mr1vTvH0s2Q&bvm=bv.1357700187,d.aWc. Accessed 15 May 2013
24. Hall C, Dale B, Pimentel D (2011) Seeking to understand the reasons for different energy return on investment (EROI). Estimates for biofuels, sustainability 2011, 3. http://www.google.ca/url?sa=t&rct=j&q=&esrc=s&source=web&cd=9&ved=0CG0QFjAI&url=http%3A%2F%2Fwww.mdpi.com%2F2071-1050%2F3%2F12%2F2413%2Fpdf&ei=670vUYLGDuSMyAHOzoD4AQ&usg=AFQjCNEmKNyJNNb8PdlA2_isXKlJQAtvUg&bvm=bv.43148975,d.aWc. Accessed 15 May 2013
25. Reed S (2013) North Africa's prospects as energy goliath are fading. New York Times. http://www.nytimes.com/2013/02/23/business/global/23iht-natgas23.html?_r=0. Accessed 25 May 2013.
26. Lambert H, Balogh P, Gupta A (2012) EROI of global energy resources: preliminary status and trends. State University of New York. http://www.dpuc.state.ct.us/DEEPEnergy.nsf/fb04ff2e3777b0b98525797c00471aef/a546c841171f7a8485257ac90053565a/$FILE/R.%20Fromer%20Attachment%20-%20EROI%20of%20Global%20Energy%20Resoruces.pdf. Accessed 26 May 2013
27. Raugei M et al (2012) The energy return on energy investment (EROI) of photovoltaics: methodology and comparisons with fossil fuel life cycles. Brookhaven National Laboratory: U.S. Department of Energy. http://www.bnl.gov/pv/files/pdf/241_Raugei_EROI_EP_revised_II_2012-03_VMF.pdf and http://www.nytimes.com/2013/02/23/business/global/23iht-natgas23.html?_r=0
28. Prieto P, Hall C (2013) Spain's photovoltaic revolution. The Energy Return on Investment. Springer, New York

Chapter 3
It is the flow, stupid!

3.1 Introduction

Imagine that you had two wealthy aunts and a wealthy uncle who all adored you and all suddenly died. Being wealthy they all decided to leave you substantial amounts of money, let us say US$ 500,000 each. Each one had a different view of your ability to manage money though and put different conditions on when you could withdraw the money and how much. Aunt 1 decides that you should get US$ 100,000 each year until the money runs out. Your uncle is a bit more careful with money and not very impressed with your money management capabilities. So, he decides that you can only withdraw US$ 25,000 per year. Aunt 2 makes your uncle look positively spendthrift, she decides you are just not old enough yet so you will have to wait 5 years before you can withdraw any money and then withdraw only US$ 25,000 per year. Overall the amount of money is the same, but the conditions on the rate at which you can withdraw it will have a real impact on your lifestyle.

Energy sources vary in the level and timing of possible flow, very much like the money gifts from your aunts and uncle. Even though the amount in various reservoirs may be exactly the same, our society relies upon the flow of energy that can be provided in a given year, not the overall amount in the ground. Unfortunately for us, the easy fossil fuel deposits that we drilled first are also the ones that provided the biggest flow. The Ghawar field in Saudi Arabia, the monster of all oil deposits, is like the first aunt on steroids. Discovered in 1948 and starting production in 1951, Ghawar has been providing 5 million barrels per day, year after year ever since. No other oil field comes close.

The Alberta tar sands may contain more oil than Ghawar and all the other Saudi oil fields combined, but it has taken decades just to get production to near 2 million barrels per day. Extraction of oil from these tar sands is a huge, phenomenally complex, and capital intensive job. Even the optimists see this source growing to only 3.5 million barrels per day by 2025. The tar sands are a bit like Aunt 2, but even more so. Drilling in the deep ocean, and now even in the Arctic, is also very much like Aunt 2. You have to wait a long time to get the oil flowing and that flow is quite limited. Many of the other energy sources, such as nuclear, wind, and solar, also have limitations on the flow of energy that they can provide. During the

R. Boyd, *Energy and the Financial System,* SpringerBriefs in Energy,
DOI 10.1007/978-3-319-04238-1_3, © Roger Boyd 2013

construction phase no energy is provided but a lot of energy is used and a lot of money invested. When the system goes online, the energy flow is like an annuity through regular instalments.

This rate-of-flow concept also applies to other aspects of energy production. For example, climate change may significantly reduce the productivity of many existing hydro plants and restrict the flow of energy from hydroelectric facilities. A warmer climate will increase the rate of evaporation from a hydro dam's reservoir, and in many areas, reduce the precipitation that supplies the dam. In winter, much of the precipitation arrives as snow which acts as a natural storage device to balance out the water flows between the seasons. This natural storage capacity dwarfs that of the dams downstream. If the snow is replaced by rain and immediately flows into the rivers, dams may be unable to hold the extra flow and will be forced to let it flow downstream. With less precipitation in the summer months, and no release of melted snow, the output of the hydro plant will be curtailed, perhaps severely. In some areas the retreat of glaciers will impact the hydroelectric output. At first, when they are melting faster, the flow of water will increase, but once they are severely or completely diminished, the summer flow will be greatly reduced or, possibly, disappear. Both nuclear and fossil fuel electricity-generating plants rely upon access to large amounts of water for the cooling and steaming of the generators, hence, the location of many next to rivers and lakes. A drop in water levels may reduce the amounts available to the generating plants, thus, reducing their safe operating levels. An increase in water temperatures may also have the same impact.

Higher temperatures and lower precipitation may also affect the crop yields upon which projections of bio-fuel production are based. Thus, climate change may significantly restrict the production of biofuels. Climate change also seems to be bringing more chaotic weather, which makes the growing season highly unpredictable and increases the extremes. Neither of these effects will be good for crop yields. Such effects will be most seen in the sub-tropics, which include the southern USA, Australia, Argentina, the Mediterranean, northern India and the Himalayas, and more than half of China.

Then, there is the sudden rise of hydraulic fracturing, "fracking," to exploit deposits of otherwise unrecoverable oil and gas. The return is almost immediate, but the flow is rather measly, a bit like the tight-fisted uncle. The Barnett shale is a good example. There is a lot of oil and gas in the shale but each well that gets drilled does not get a lot of flow, and that flow rapidly decreases over time. Only by repeatedly "re-fracking" wells and drilling hundreds of wells at the same time can a respectable production rate be maintained. Ultimately, for shale, the amount of flow rapidly increases at the beginning as hundreds and even thousands of wells are drilled, but as the high well-depletion rates kick in, more and more wells have to be drilled to offset the decline with new output.

As a society, we have been living off many generous aunts and uncles for a very long time, with the majority of energy production relying on large and easily recovered fossil fuel deposits found decades ago. Those will all peak at some point and their production will decline as has already happened with the Cantarell field in Mexico, the Alaskan North Slope, and the North Sea. All of these fell rapidly from their peak to become shadows of their former selves within a decade. With only the most parsimonious uncles and aunts left as new providers, the flow of energy available to us will drop even as the headlines celebrate the huge deposits of tar sands, heavy oil, shale oil and gas, deeper and deeper ocean wells, and even arctic drilling. With more of the current and future flows being utilized in the production of new energy due to decreasing net energy levels, the negative effect on society will be intensified. The resulting lower flow of net energy will lead to a lower standard of living [1].

3.2 Oil: Probably Already Peaked, Declining Soon?

As oil powers the vast majority of global transportation, a reduction in the amount available will have very serious repercussions. I still have vivid memories of driving down the English M1 motorway in 2000, during a successful blockade of oil depots by truck drivers, praying to find a gas station without an endless line of desperate motorists. Exacerbating the situation, "just in time" inventory practices had limited many of the local stocks of food. Within days, store shelves were emptying and threatening even more desperation. Without government intervention to remedy the situation, the UK would have rapidly ground to a halt, with an increasingly hungry and restive population. The ability to develop new sources of renewable and fossil fuel energy would also be greatly impacted as oil is used at every stage of the process from the mining of the required raw materials to the transportation of parts, to the implementation site, to the ongoing maintenance.

Since 2005, oil production has been running, on average, around 74 million barrels per day, with new production flows only just offsetting the depletion and reduced flows from current wells. At some point, which may be very near, the relatively small flows from the new production fields will not be able to offset the decline of the older much more generous fields, and the daily flow of oil will start to decline. Global oil production has become much like Lewis Carroll's Red Queen, running faster just to stay in the same place. Even now, the current production plateau may be masking a decline in the net energy provided from oil production, as increases in oil volumes from tar sands, deep water, and shale just balance the depletion of higher net energy conventional oil fields. We may be getting the barrels, but we are not getting the same amount of net energy they used to deliver. Shale oil has been touted as a huge new source of energy, facilitating "energy independence" for the USA. It has allowed the US oil production to increase, but at the global level, has only aided in offsetting the depletion of current wells. The long term viability of shale oil production has been questioned given the relatively low per well production and very high depletion rates. Some researchers have also pointed to the relatively small "sweet spots" within shale oil areas, with well productivity falling off rapidly with an increase in the distance from the sweet spot [2].

A few years ago, Jeffrey Brown and Samuel Foucher focused on the demand for oil within the oil exporting countries, calling this view the "Export Land Model"

[3]. As oil-exporting nations increase in population and become wealthier, domestic demand for oil grows and the balance left for export falls. Adding to the problem, there is a tendency for such nations to keep energy costs low to either support the local economy or buttress unelected elites, as in many of the Middle East countries. This only increases the amount of their energy production used domestically, as low prices foster demand and inefficient usage. Thus the peak of "internationally available oil" will happen sooner than the global production peak. The fall from that peak may be accelerated as the oil exporters may decide to reduce production of their scarce and increasingly valuable oil resources to stretch out the period of required transition. Already, Indonesia and Egypt have moved from being oil exporters to oil importers and Mexico, Malaysia, and Iran are seeing oil exports fall faster than their overall production levels. This is not good news for countries that rely heavily on oil imports like the USA, China, Japan, and Europe.

To make things even worse, an increasing amount of world oil exports are coming from countries which are less stable than the more established suppliers. Some very interesting countries have been added to the list of significant oil exporters, such as Nigeria, Angola, and Algeria. When you think of these three countries what kind of images come to mind—images of peace, lawfulness, honesty and happiness? Or do images of scary people with guns, civil wars, political activists being executed, or widespread graft and outright theft by the powerful come to mind? Nigeria, with a production of over 2 million barrels per day, is now the eighth biggest oil exporter in the world compared with places like the USA and Europe. Close behind in oil production volume is the country of Angola. Older readers may remember the initials MPLA (Movement for the Liberation of Angola), FNLA (National Front for the Liberation of Angola), and UNITA (National Union for the Total Independence of Angola), three groups that fought a very nasty civil war in Angola over nearly three decades. Not far behind is Algeria, which only lifted a state of emergency in 2011 and where security is so lacking that at the beginning of 2013 a natural gas facility was attacked and several foreign workers killed. Another problematic country is Libya, where a stable dictatorship has been replaced by tribal warfare and ongoing disruptions to oil exports. So, if the inhabitants of the oil-exporting countries are not consuming more of their oil themselves, their exports may be affected by some civil disturbance which shuts down oil production. Perhaps that is why the US armed forces have formed an "African Command" and the European countries were so quick to engage with the Arab Spring. At least, increasingly unstable Egypt is no longer a worry; it has become a net oil importer. Syria is not far behind, with its oil exports declining well before the current civil war started.

Other options for producing oil such as coal to liquids (CTL), gas to liquids (GTL), and plant-based ethanol are all very limited in their ability to increase "oil" flow to offset declines in both conventional and unconventional oil production. CTL and GTL facilities are extremely complex undertakings which are open to large cost overruns. A good example is the Pearl GTL project in Qatar where costs have nearly quadrupled during the construction of the facility. Ethanol production competes with the production of food, for fertile land; and as seen in the US farm belt in 2012, crop yields can be badly affected by weather extremes. Another possibility is liquefied natural gas (LNG), which can be used as a transport fuel, especially in larger vehicles such as buses, trucks, and ships. However, refitting or replacing millions of oil-dependent vehicles would require major investments and take years, if not decades. With peak oil already a present reality, it may be too late for a transfer to natural gas to offset falling oil production. In addition, such a move would significantly increase demand for natural gas, thus, hastening its own production peak.

The economic and social pain caused by these reduced oil exports will not be felt evenly between importing countries. Some importers may be able to outbid others, forcing an inordinate share of export declines onto a subset of unfortunate importers. In this respect, the newly developing countries may have an advantage over the more highly developed nations. The latter have already used the highest pay-off, low-hanging-fruit uses of oil, while developing countries such as China are still working through those higher pay-off applications. With a higher economic return for each incremental barrel of oil, the developing countries may be able to outbid the already developed countries. This could possibly speed up the convergence between developed and developing countries, as the former are forced into ongoing recession due to oil shortages and the latter continue to grow. Interestingly, the US oil usage has declined in the past few years, even during the economic recovery, with the US gasoline usage hitting a 16-year low in 2012. Both Japanese and European oil consumptions have also been falling since the start of the 2008 recession, especially in the economies most affected such as Portugal, Ireland, Italy, Greece, and Spain. These consumption reductions by the USA, Europe, and Japan have allowed countries such as China to increase oil imports even while global oil exports declined (Fig. 3.1). The UK, which has depleted its coal reserves, and its natural gas and oil reserves to the point where it is a net importer of both, may be one of the most negatively affected countries.

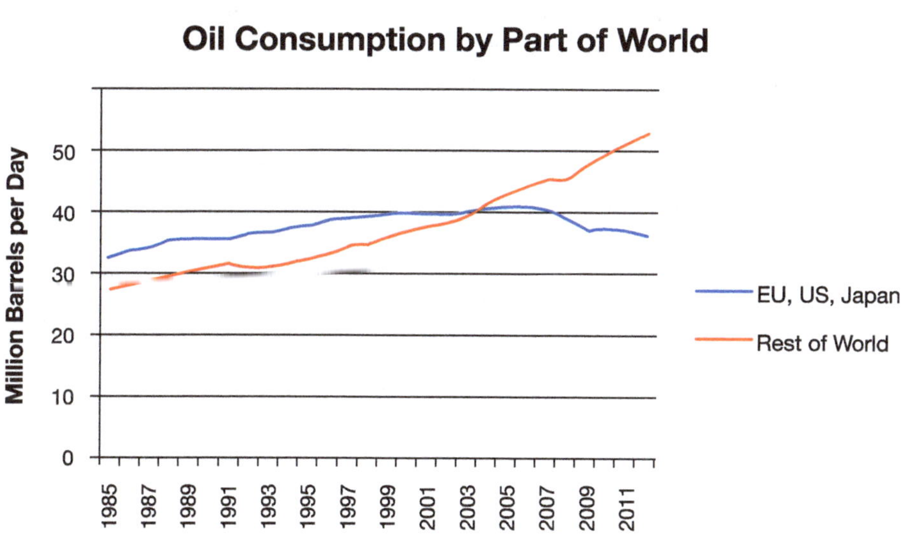

Fig. 3.1 Oil Consumption: EU, US, Japan and Rest of World. (Source: Peak Oil Demand Is Already A Huge Problem. Gail Tverberg, Our Finite World website article, November 15th, 2011)

3.3 Natural Gas: More Flow Possible and Some Substitution for Oil as a Transport Fuel

Unlike oil, natural gas supply is a predominantly regional energy source. Even with the recent increases in Liquefied Natural Gas facilities and tankers about 90 % of NG supply is still delivered through pipelines. The exploitation of natural gas fields had to wait for the pipeline infrastructure to be put in place in the post-WWII period, thus, a peak in overall natural gas production is at least a decade away.

Although conventional natural gas production has peaked in North America, shale gas production has more than offset conventional well depletion in the past few years. As with shale oil though, the long-term viability of shale gas production has been questioned given low per well production, high depletion rates, and the relatively small "sweet spots" within shale gas areas. Some of the earlier shale gas plays do seem to have peaked, or be nearing peak [4], supporting those that see shale gas as a short-term palliative rather than a long-term replacement for North American gas supplies. If this view is true, the USA may again become a net gas importer before 2020. The US production levels over the next few years will demonstrate whether such an outcome is likely. The increased production volumes may also be somewhat misleading given that the net energy provided by shale gas is significantly lower than that for conventional gas. Thus, the gas production numbers may overstate the increase in available net energy.

The lack of the infrastructure to export gas from North America, together with US government restrictions on gas exports, has meant that the increased shale gas production has lead to a regional glut and low prices. In many cases, shale gas production is not profitable at current regional prices. The lower prices have lead to a displacement of coal for electricity generation, as many plants can burn either coal or natural gas, and new natural gas capable plants are being added. Shale gas production has also displaced the expected the US LNG imports, and thus, reduced the global competition for such supplies.

European conventional natural gas production has, or will have very shortly, peaked. Large-scale imports from Russia, together with supplies from North Africa, have become a significant part of the European energy supplies. LNG supplies from the Middle East are also increasing. Shale gas production is now being explored as a way to reduce dependence on these external supplies, but is at a much earlier stage of development than North America and costs will also be higher due to more dense populations, more difficult geology, and stronger environmental legislation. In addition, the extensive energy extraction infrastructure that could be easily utilized within the USA is not present in the European countries. Thus, shale gas may have much less of an impact than in North America.

As with oil, the Middle East and northern Africa have very large conventional natural gas reserves and are significant providers of LNG, mostly to the Asian markets. Four Middle Eastern countries, Iran (16 %), Qatar (14 %), Saudi Arabia (4 %) and the UAE (3 %) have over a third of global deposits and Qatar is the world's largest supplier of LNG [2]. However, the "Export Land Model" is also applicable to natural gas exports, with production increases in Iran, Saudi Arabia, and the UAE being used predominantly to serve the domestic market [5]. The balance between increased local demand and production levels will dictate the ability of the region to increase exports.

Central Asia has become a highly promising source of natural gas exports. Substantial reserves have been discovered, facilitating production which is far in excess of local demand. A gas pipeline to China has been completed and a connection to Europe is scheduled to be completed in 2017 [6]. Gas exports from the region will double during the next decade with China taking about half of all exports [5]. Such exports will become more critical as Indonesia and Malaysia switch from being net exporters to net importers in the next few years. Increases in Australian exports will also help fill the gap left by Indonesia and Malaysia. The rapidly growing economies of India and China, together with a Japan without its nuclear power, will require increasing levels of LNG supplies. China and Australia also have significant shale gas resources that have yet to be developed. With an authoritarian government that could possibly force through shale gas development in the former and low population densities in the latter, both countries may be successful in exploiting these resources.

Producing half as much carbon dioxide produced when burnt, natural gas has been touted as a more climate friendly fuel than coal. Unfortunately recent research points to a high level of leakage from natural gas production sites [7]. The majority of natural gas is methane which has approximately 20 times the heat-trapping ability of carbon dioxide. Methane stays in the atmosphere for a much shorter time than carbon dioxide, thus its heat-trapping effects are condensed into many fewer years. In those years, its heat-trapping ability can be as high as 100 times that of carbon dioxide. Thus, the careless or unregulated substitution of natural gas for oil and coal may actually speed up climate change in the short run.

In the next 10–15 years, significant increases in global natural gas production are possible. However, the net energy of many of the newer supplies is significantly less than the older fields so increases in net energy will lag increases in gross production volumes. LNG can also be used as a substitute for oil as a transport fuel, especially for buses, trucks, and ships. Royal Dutch Shell is chartering two river barges which are designed to run on LNG, and is also planning to provide LNG fuel stops along a truck route in Alberta, Canada [8]. If such uses become widespread, they may soften the blow of falling oil production, but greater gas usage will certainly speed up the depletion of natural gas reserves. With the increasing impacts of climate change any increases in production may also be limited by external constraints put in place to limit the amount of heat-trapping gases emitted into the atmosphere. The very high short-term climate impact of methane, together with environmental issues affecting shale gas production, may mean that natural gas production comes under the same level of citizen resistance currently impacting coal production.

3.4 Coal: The Old Workhorse Still has Something Left

Coal was the first fossil fuel to be heavily exploited and the levels of its depletion mirror the timing of individual country's industrialization. The UK, as the first to industrialize, has burnt through its huge deposits of coal and now has a near-trivial level of coal production. At its peak, at the beginning of the twentieth century, UK coal production employed well over a million miners; now there are only about 10,000. Coal production in all of Western Europe is much reduced from its peak, having fallen 37 % between 1981 and 2010 [9]. A majority of the remaining production is heavily dependent upon state subsidies. In the former Soviet Union coal production has also declined, falling 31 % between 1980 and 2010. But even at this reduced production level, Russia is still the third biggest exporter of coal globally. With the planned removal of European subsidies by 2018, a significant portion of the remaining coal mines in Western Europe will be unprofitable and have to close. Since its coal production is withering away, Europe finds itself in the precarious position of becoming dependent on imports of coal from distant locations. Europe's coal situation mirrors its high reliance on imported oil and gas, making the continent very vulnerable to any shortfalls in available fossil fuels.

North America was one of the early industrialized regions, but it had extremely large coal deposits and a much lower population density than Europe. Thus, the US coal production has not yet peaked, with production now double what it was in the 1920s [10], when the UK had already reached its production peak. The increasing use of natural gas for electricity production has reduced regional demand for coal, but this is being somewhat offset by increased exports. The US is the world's fourth biggest coal exporter. Central and South America, together with the Middle East, have only small coal deposits. The vast majority of South American deposits are in Colombia, where production has expanded rapidly in the past few decades making it the fifth biggest global coal exporter.

While there is significant opposition to the building of new coal-fired power stations in the wealthier countries, the developing countries are rapidly building such power stations. The rapid growth of Asia as it rushes to catch up with the earlier industrialized nations has been predominantly fuelled with coal. The region's coal production doubled between 1980 and 2000, then more than doubled again between 2001 and 2010 [9], accounting for half of global coal production in 2010. China has grown its coal production at an even faster pace utilizing its extensive coal reserves to power industrialization for its 1.3 billion plus population. The plans for Chinese coal production envisage a rate of 3.9 billion tons by 2015, which will be about four times that of the USA and nearly half of global coal production. India, with a population of 1.27 billion, is also rapidly increasing coal production with a 400 % increase between 1980 and 2010. Even with such rapid growth in local coal production, China and India have both increased coal imports, driving coal production increases in Indonesia, Australia, and South Africa (the world's first, second, and sixth biggest coal exporters, respectively). Japan and Korea, with little local coal production, are also heavily dependent upon imports. Japan's coal dependence has increased with the closure of its nuclear facilities in the aftermath of the Fukushima crisis.

The very rapid increases in production in Asia and the coal exporting countries have raised global coal production by 91 % between 1981 and 2010 [10]. With oil production on a flat plateau at best since 2005, ongoing growth in global energy supplies and thus GDP has been provided by increases in coal, and natural gas, production. In fact, the share of global energy supplied by coal is set to exceed that of oil once again. Many commentators believe that continued increases in coal production can be maintained for decades. Others, such as Heinberg [11], have questioned the validity of coal reserve amounts, and see a much earlier peak in coal production. A significant increase in coal production may be able to offset the ongoing reduction in oil supplies. However, much of the increased coal production will come from relatively new areas which are quite distant from coal consumers which may limit their use. In the USA, vast western coal reserves are situated far from the main consumers in the east and thus extensive (and expensive) new railroad investments will be required to support any significant increase in production. The same can be said for the large coal reserves in Mongolia, Siberia, and Alberta, Canada. Further, these new sources are often lesser coal grades which produce less energy. Thus, the growth in volume from the new sources of coal may not yield an equal growth in net energy delivered.

Oil is predominantly used as a transportation fuel, while coal is used primarily to produce electricity. For coal to replace oil either a significant amount of the world's transportation system would have to be converted from oil to electricity or coal would have to be used to produce oil through CTL technology. Large scale energy transitions have tended to take decades, and the replacement or retro-fitting of a significant number of gasoline and diesel powered cars, trucks, and trains to use electricity would certainly be a long and costly process. The CTL process is highly energy intensive and the Energy Return on Investment (EROI) of the resulting product is only about 4:1, well below that required for advanced industrial societies. Changing the production and consumption infrastructures to allow for a large-scale conversion from oil to coal-fired electricity, together with the low EROI for the CTL process, argue against any ability for coal to replace oil in the foreseeable future.

Beyond this, coal may still end up being the biggest fossil fuel provider of energy for many decades to come, and provide some stability as other energy sources deplete. For regions that have depleted their coal reserves and have little remaining natural gas or oil there will be very serious consequences unless they can successfully develop alternative sources. Many European countries fit this unfortunate profile. India and China will benefit from the increased exploitation of their coal reserves, with China being the major player. A peak in Chinese coal production due to geologic limitations may very well define the global peak for coal. China's increasingly prosperous citizens are also becoming much more active in pushing for better air quality, which may force a lower peak in coal production. North America, Australia, and Russia will also benefit from their significant remaining coal deposits. The world as a whole may be faced with the difficult choice of either burning more coal to keep something like the current civilization going or burn less coal to limit global warming and accept some level of societal simplification, and possibly, collapse. A complicating factor is that the sulfur pollution produced from burning coal acts to block sunlight and offsets some of the warming effects of greenhouse gases. This has been referred to as "global dimming," and it has been proposed that this dimming effect has kept temperatures 1–2 degrees cooler than they otherwise would be [12]. The sulfur pollution created by burning coal only stays in the atmosphere for a short time while the carbon dioxide produced stays for many decades. Thus, recent large increases in coal usage may have increased the dimming effect somewhat mitigating the warming trend; so a significant reduction of coal burning may produce a rapid jump in temperature.

3.5 Hydroelectricity: Small Increases Possible

As discussed in the previous chapter, it may be possible to double the output of hydroelectricity globally. With most of the best sites in the advanced industrial countries taken, it will be the industrializing countries that will benefit from added hydroelectric output. One exception is Canada which has large untapped hydro potential, and already gets 59 % of its electricity from hydro [13]. The construction of new hydro plants may be greatly offset though by the effects of climate change, which has the ability to greatly decrease hydro output in the American Southwest, India, Pakistan, China, and parts of South America. A study of possible climate change effects on the US hydro output envisages the water levels behind the Glen Canyon Dam, and the Hoover Dam possibly dropping to zero by mid-century [13]. Large hydro-electric dams also take a long time and a lot of energy to go through the planning, environmental review and construction phases. So, despite their potential, for a number of years they create a net energy drain before being able to provide a positive flow of energy.

In the short term, building more dams may actually exacerbate climate change. The construction of new hydro plants adds to climate change emissions through the energy used to build them and the decomposition of the vegetation flooded by the rising waters. With the lush vegetation of many of the possible undeveloped sites, the effect of this decay may be significant [14]. For existing dams these early emission costs have already been paid and they now offer very low emissions electricity generation.

3.6 Nuclear: New Capacity Significantly Offset with Retirements

Many of the current nuclear power installations were constructed decades ago and are reaching the end of their life cycle. The useful life of a nuclear installation is about 40 years. Nearly 45 % of nuclear plants worldwide are over 30 years old; over

90% have been operating for over 20 years [15]. Some have had their useful life extended through renovations and the loosening of regulatory standards, but such extensions are limited. With such a large number of ageing nuclear power stations, construction of new nuclear generating capacity may only balance the retirement of current plants. To maintain this balance it has been estimated that approximately 20 new nuclear power installations will be needed per year to offset the removal of aging ones [15]. So, despite what promises to be a vast investment, the amount of nuclear power produced globally may remain relatively stable overall.

Nuclear power stations have a number of vulnerabilities to both weather and geologic events. First, they need to be located next to water sources such as the sea coasts or along rivers. This means that floods and coastal storms can render them inoperable. This was not only seen at Fukushima following the Japanese earthquake, but also at the Fort Calhoun facility near Omaha, Nebraska. Photographs of the Fort Calhoun plant surrounded by the flooding Missouri river and protected by a giant "rubber barrier" vividly illustrated the risk of such floods [16]. Two years later, in early 2013, the plant had still not re-entered service. In 2012 Hurricane Sandy underlined the vulnerability of coastal nuclear facilities. Many were shut down for safety reasons as Sandy approached. Climate Change is destined to bring more intense rain events, raise sea levels and intensify storms. Compounding the flood and storm threat, the legacy of Three Mile Island, Chernobyl and Fukushima has been to increase mandated levels of safety infrastructure, and embolden communities to resist their "nuclear neighbors" even more than before. The much-vaunted US "nuclear renaissance" is being stymied by both citizen and regulatory resistance. This is especially true of the nuclear plants built in earthquake zones such as Diablo Canyon and San Onofre in California, and the Indian Point facility in New York State. Even if they are not hit by an earthquake, local citizen opposition based on fear of such events may curtail their active lives.

Recently, slow growth in electricity demand together with the low regional cost of natural gas has rendered nuclear plants uncompetitive. The first authorization for a new US nuclear plant in more than 30 years, made over the dissenting voice of the chairman of the Nuclear Regulatory Commission, was issued in February 2012. Officials now estimate that only five new reactors will reach service by 2020 [17]. Within a year after its authorization, the new plant had run into construction problems and may be falling years behind schedule, making even the forecast of five new US plants by 2020 look quite optimistic [18]. In the wake of Fukushima, both Germany and Japan committed to a phased shutdown their nuclear electricity plants and the Canadian province of Quebec decided to close its ageing single nuclear plant. The effect of such policy driven retirements, together with the closure of aging plants, will be predominantly felt in the advanced industrialized regions. The benefits of adding nuclear capacity will accrue to such countries as China and India who have few plants currently, but have plans for many in the future. A lack of democratic structures will facilitate the implementation of such plans, especially in China where 26 nuclear plants are under construction to add to the 15 relatively new ones already operating. Many more are planned. Currently, India has seven plants under construction [19].

Uranium is also a depleting resource. Nearly half of the demand in the past couple of decades was filled from existing uranium stockpiles or by the retirement of nuclear weapons as part of the 1993 "Megatons to Megawatts" deal between the USA and Russia. Just as with the fossil fuels the easier deposits of uranium were mined first; newer mines have lower concentrations of uranium and higher costs. Canada is the only country with untapped ore grades of 1 % (from which the mining of 100 pounds of ore yields 1 pound of uranium), the majority of the rest of the world has ore grades below 0.1 %, with two thirds of that being below 0.06 % [13]. Between the exhaustion of stockpiles and reductions in supplies from weapons decommissioning, it has been estimated that uranium mining output will have to increase by about 50 % to continue supplying *current* demand. Taking into account planned growth it has been estimated that economically viable uranium deposits could be depleted within 30 years [14]. Investment in the uranium mining industry went through a two decade slump from which it has only recovered in the last few years. Given the need to increase output by at least 50 % to meet both continuing demand together with increased demands should nuclear usage grow, at least a doubling of uranium mining may be required. However, the opening of a new uranium mine can be a long and arduous process. One recent project, Cameco's Cigar Lake mine has been delayed by over 5 years due to repeated flooding [20].

After being used to generate electricity spent nuclear fuel can be reprocessed to recover usable fission materials that can be used to generate about 30 % more electricity. As with nuclear reactors, very large up-front investments are required to build reprocessing plants before output begins. Reprocessing also requires uranium prices as high as US$ 360 per ounce to be profitable [21]. A report commissioned by the French Prime Minister in 2000 estimated that France would have saved US$ 33.5 billion by *not* reprocessing nuclear fuel [22]. Currently commercialized reprocessing technologies also produce weapons grade plutonium, a factor which has lead to significant citizen resistance and a policy decision against reprocessing by the USA. By 2018 both of the UK's reprocessing plants will have been shut down, and half of the remaining global capacity will be provided by the single La Hague facility in France [23]. Another quarter of the remaining global capacity will be supplied by the Japanese Rokkasho facility, which is expected to be cleared for production in late 2013 after well over a decade of construction [24]. Whether or not the Japanese population will allow this plant to operate is of course open to question in the aftermath of Fukushima.

Nuclear breeder reactors, which would greatly improve the efficiency of the uranium usage, comprise incredibly complex feats of engineering that have yet to be proven commercially viable. The use of thorium, which is much more abundant than uranium, has been discussed but is still only a concept. The time for such technology to go from the drawing board to commercial installations is counted in decades, and thus, does not figure meaningfully in the timeframes required to replace fossil fuels and reduce greenhouse emissions. Nuclear fusion is even farther away from being commercialized, with commercially viable plants being many decades in the future at the earliest.

3.7 Wind Could Provide a Useful Percentage of Energy Needs

Wind power produces electricity, which is forecast to be the fastest growing energy carrier over the next decades. With BP forecasting a growth rate of 2.6 % per year [25], global electricity demand will be about 30 % larger in 2022 than in 2012. The International Energy Authority has forecast an even greater increase in electricity demand over the next couple of decades [26]. Thus, any increase in electricity supply from wind will go first to meet the increased demand, and only after that has been met will any fossil fuel sources be displaced. Wind turbines are dependent on there being wind and as that is not always the case their actual output never equals the stated capacity to produce electricity. The capacity factor, the actual percentage of the capacity that will be produced over a given year, averages about 25 % globally. Thus, forecasts of installed capacity can be very misleading as a gauge of how much electricity will actually be produced by wind turbine installations.

Even after the rapid growth of wind power over the past years it still represents only about 1 % of global energy supplies in 2012. Wind energy capacity has been growing at approximately 30 % per year for the past fifteen years [26], but it would be unrealistic to expect it to keep growing at that rate. As the installed base grows adding another 30 % requires bigger and bigger additions. For example if the 30 % growth continues for the next 5 years, the installed base would quadruple in size, with three times more capacity added in the next 5 years than had been added in the previous 15. For such growth to continue for the following 5 years, twelve times more capacity would be required than that in place at the end of 2012. This is plainly unrealistic without the impetus of emergency actions taken by governments to move rapidly to non-fossil fuel energy sources. Some of the previous growth has also been dependent upon government subsidies, but many austerity enthralled governments are starting to reduce such subsidies. In addition, as the scale of wind implementations grows the cost of government incentives will grow in parallel creating a greater and greater pressure to reduce such subsidies.

Even wind energy supporters such as the Global Wind Energy Council and Greenpeace, using their more optimistic scenarios, assume future growth rates of the wind power sector will start at about 20 %, and then decline to 6 % by 2030. The resulting share of electricity supply from wind energy, taking into account increases in demand, will be between 14 % and 25 % by 2030 for the two most optimistic scenarios. Even at the high estimate, all of the increase in output would be taken by increased demand rather than the replacement of fossil fuel sources. If the dreams of millions of electric cars were to be realized increases in electricity demand would be much greater, reducing the overall share of wind energy and the possibility of fossil fuels ultimately being displaced by it.

3.8 Bio-Fuels: From Really Small to Small

Bio-fuels provide currently about 0.5 % of global energy supplies, with two countries, the USA (corn ethanol) and Brazil (sugar cane ethanol) providing almost 90 % of global production. The production of corn ethanol in the USA relies upon government subsidies and mandated increases in the amount of bio-fuels blended into the US fuel supply through 2022. Net energy assessments for corn ethanol fall within a tight range above and below zero, supporting the view that without government subsidies and mandates corn ethanol production would simply disappear. About 40 % of the US corn production is currently used to provide about 10 % of the US gasoline consumption [27]. At the societal level this makes absolutely no sense, and points to the ability of politically powerful economic entities to take advantage of the drive for alternative fuels to enrich themselves. Such "corporate welfare" redirects limited resources away from more fruitful areas. Further increases in the amount of ethanol blended with petroleum are also limited by the ability of car engines to deal with higher and higher amounts of ethanol. This "blend wall" is close to being reached for standard gasoline car engines. The recent drought conditions in many growing areas have also intensified concerns that the redirection of crops from the food supply will contribute to food shortages and rising prices worldwide. With climate change predicted to intensify such drought conditions, the ongoing use of large percentages of crop production for fuel may become politically untenable. Given the low, or even negative, net energy of corn ethanol any increases in production will not meaningfully affect the net energy available to society.

Brazilian sugar cane ethanol production provides much higher levels of net energy, with an EROI of about 7:1 [28]. Over 90 % of all cars and light vehicles in the country are flex-fuel models that can run on any mix of gasoline and ethanol, removing the blend-wall as a limiting factor. Mario et al., have identified a number of factors specific to the growing areas in Brazil which account for a significant part of the net energy from sugar cane ethanol [28]. Brazilian cane varieties seem to need much less fertilizer than others and the high rainfall leads to a very low requirement for irrigation. Also, the bagasse byproduct (fibrous matter left after the cane juice is extracted) can be used to supply much of the electricity and heat required to convert sugar cane into ethanol. Relative to its population, Brazil has an extremely large amount of available agricultural land that can be utilized for sugar cane production. Only a few tropical countries around the world have this mix of extensive available land and climate factors, meaning that the Brazilian success will not be widely replicable. In the future much of Brazil's agricultural land may be significantly affected by climate change primarily through reduced rainfall, higher temperatures, and less predictable weather. These factors would increase the need for energy intensive irrigation and reduce crop yields, driving down the net energy supplied. Recently, 2 years of bad harvests in Brazil, combined with a lack of new investment due to economic recession, have lead to the increased consumption of gasoline and ethanol imports from the USA [29]. This underlines one of the fundamental differences between bio-fuels and fossil fuels; from year to year, the supply of the latter is much more predictable than the former. The output of an oil well is seldom affected by changes in temperature or precipitation.

Cellulosic ethanol has been put forward as the more productive "next generation" of ethanol production, with estimates of EROI range as high as 18:1 [30], although there are no commercial plants operating currently. The technology required is at a very early stage and is at least a decade or more away from being commercially provided in significant amounts [29, 31]. The production of fuel from algae has also been discussed, but it is also at a very early stage of development. Neither of these will have a significant impact within the foreseeable future, and in fact cellulosic ethanol production is already well behind the assumptions made in the US ethanol mandates.

A fundamental issue with the use of plants to replace significant amounts of fossil fuels with bio-fuels is the assumption that the earth's ecosystems have large amounts of unused biomass available for conversion into fuel. Nature does not produce waste; instead biomass serves as food to the animal and insect population and as nutrients for the soil either directly through death and decomposition or indirectly when excreted by animals and insects. It was estimated that in 2005 that humanity already utilizes about 25 % of the earth's Net Primary Productivity (NPP), the amount of plant material produced on Earth. This usage has continued to grow, and with current trends in population and per capita consumption, could exceed 50 % by mid century without any additional usage for fuel [32]. If large additional amounts of plant material are to be used to fuel human transportation humanity may require the overwhelming majority of Earth's NPP. This can only happen if increasing amounts of plant material are taken away from their current consumers, resulting in accelerated species loss and soil degradation. The effects of such large scale increases in humanity's call on Earth's incredibly complex ecological processes are unknown, but are likely to prove quite negative.

The International Energy Authority (IEA) understands the ecological limits to the use of bio-fuels, stating that the usage of agricultural and forestry residues will be dependent upon "the sustainable portion that must remain in the field to replenish soils and maintain future crop yields" [33]. The IEA estimates that bio-fuels will grow from 3 % of world transport fuels to 8 % in 2035. With net energy rates significantly below that of the fossil fuels, projected volume increases significantly overstate the amount of new net energy that could be provided to society. Fossil fuels represent the product of millions of years of photosynthesis compacted into incredibly energy dense forms, which humanity is using up at a rate thousands of times faster than they were formed. With bio-fuels only the current product of photosynthesis is available, and thus they will never be able to match the amounts of energy made available through the formation of fossil fuels. Large-scale attempts to do so will only accelerate the degradation of the global ecosystem upon which humanity relies for its survival.

3.9 Solar: From Tiny to Really Small

Solar photovoltaic energy (PV) has grown extremely rapidly over the past decade, but from a very small base. Between 2000 and 2011, it grew from 1.5 GW of installed capacity to 67 GW, averaging a 50 % annual growth rate during the last 5 years of that period [33]. This capacity is heavily concentrated in European countries, with the European Union representing over 75 % of the installed capacity. Even using the optimistic uninterrupted rapid growth scenario of the IEA, the installed capacity of solar PV is forecast to increase to only 600 GW in 2035 which would account for about 2 % of global electricity generation. All of this growth would be far less than the growth in electricity demand during the forecast period, meaning solar would not replace any current fossil fuel energy sources. Concentrated solar power (CSP) has a much smaller presence than solar PV, with only 1.3 GW of capacity installed in 2010. Again, using the IEA growth scenario, solar CSP would grow to 72 GW by 2035 which is approximately equal to amount of solar PV installed in 2011! This would constitute a negligible share of the global electricity supply.

When looking at solar energy, as with wind energy, installed capacity can be very misleading as it represents the maximum possible output rather than the average. The amount of solar energy available is affected by the time of day, physical location, local weather, and technology. The capacity factor can vary from under 20 % to over 30 % for solar PV and solar CSP without storage. Storage can raise the solar CSP capacity factor as high as 50 % [34]. The highest factors will be gained at locations close to the equator with a high percentage of clear days. The drawback of such locations is that they may be far away from the consuming populations and, if so, would require expensive long distance high voltage power lines—another vast net energy draining investment.

3.10 Wave: Really Tiny to Perhaps Tiny

An analysis produced for the UK Carbon Trust [35] estimated that approximately one tenth of current UK electricity consumption [36] could be extracted from UK waters. Given that the UK is situated on two islands surrounded by extensive continental shelves, it should provide one of the most promising opportunities for wave power. However, the uncertainties and risks of the large-scale engineering works required in water, hundreds of meters deep and about a hundred kilometers offshore, are very high. The obvious lack of any large-scale operating wave-power implementations underlines the number of unknowns, including high probabilities of time and cost escalation, together with output underperformance. With the possible exception of very specific cases, wave energy will not play a significant part in the global energy supply within the foreseeable future.

3.11 Geo-Thermal: Really Tiny, Unless You Live in Iceland

As with wave energy, except for very specific cases, such as Iceland, geo-thermal energy will not play a significant part in global energy supply within the foreseeable future. The population of Iceland, which derives a major share of its energy from geothermal sources, is only about 350,000, a proverbial drop in the bucket when compared to the global human population.

3.12 Summary

Total world energy demand in 2010 was about 523 quadrillion (that is 523 thousand billion) British Thermal Units (BTU), with a BTU being the amount of energy needed to heat one pound of water by 1 °F [37]. In the past century, this energy demand has grown by a factor of ten [37]. Global energy consumption is estimated to keep growing at about 1.6 % for the next 20 years, resulting in a 36 % growth during that period [38]. Nearly all of the growth is assumed to be in the low and medium income economies, such as China, India and Brazil (Fig. 3.2).

Fossil fuels currently provide about 87 % of current energy supplies, dwarfing the 2 % provided by the new renewables such as wind, solar, and bio-fuel. Even with relatively optimistic assumptions about the growth of such renewables, it is obvious that they will not be able to meet even the forecasted increase in energy demand, let alone replace current fossil fuel energy sources. Hydro and nuclear will be able to provide some additions, but even with their contribution, fossil fuel use will need to "increase" to meet the forecast demand. This is the elephant in the room that tends to be ignored when the outlook for fossil fuel use is discussed. As long as economic growth continues, which historically requires increases in energy use, renewable and nuclear energy will not realistically displace fossil fuel use. With both the depletion of currently producing fields, and the need to constrain fossil fuel use to limit climate change, that economic growth begins to sound impossible.

Natural gas has been touted as a climate-friendly "bridge" fuel, allowing more time for the move to non-fossil fuel energy sources, as its combustion provides much less carbon dioxide than that of coal and oil. Unfortunately this claim may be offset by methane leakages and natural gas may be just as bad, or even worse, than coal and oil for climate change. Natural gas contains mostly methane, which traps up to 100 times more heat than carbon dioxide during the years it stays in the atmosphere. Significant methane leaks have been found at shale gas sites, as noted earlier in this chapter. Far from the wells, hundreds of such leaks were also found in an analysis of the New York City area [39]. The scale of the leaks could easily trump the lower level of carbon dioxide produced when natural gas is used to generate energy.

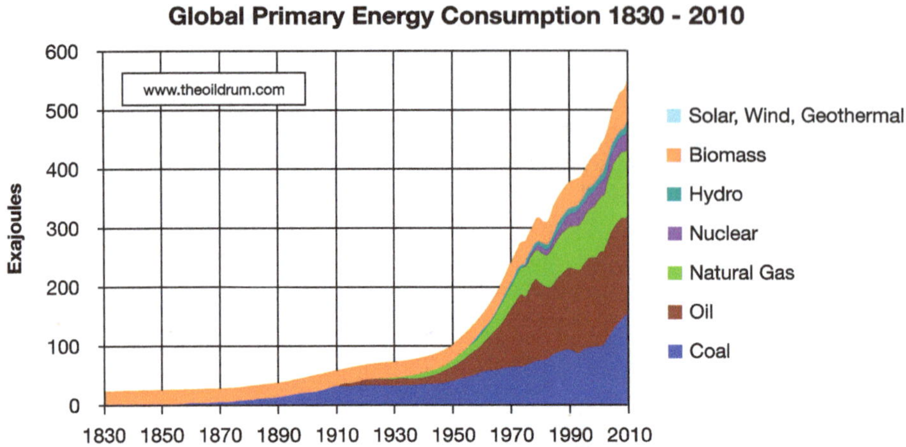

Fig. 3.2 Global energy consumption 1830–2010. (Source: Rembrandt Kopelaar, taken from The Oil Drum, accessed at http://www.theoildrum.com/node/8936)

With rising living standards in many of the oil- and gas-producing countries, their energy production will increasingly be used for their own needs first with less left for export. Thus, global exports of energy may decline, even if overall energy production is static. This trend seems to have already begun with respect to oil. Global production has been flat but the amount available for export has declined by about 2 million barrels per day between 2006 and 2011 [40]. This will exacerbate the problems for some of the energy import dependent nations, as they are priced out of the market for available exports by the developing nations which see larger marginal economic gains on their imports. This should mean that countries such as China and India can outbid richer counties for fossil fuel supplies. The tying up of oil and gas exports through the placement of pipelines, and long-term purchase agreements, predominantly by China, will also remove access to a significant amount of global exports from the wealthier energy importing nations. For example, China has continued to increase its oil consumption while the USA has reduced its rate of oil use [41]. These factors taken together may mean that the developing nations may shoulder much less of the impact of declining energy supplies in the short term. For the future course of climate change, events in China, India, and Brazil will be much more important globally than the amount of German electricity produced from wind farms.

Another of the impending limits on energy supplies will be the silting up of older hydroelectric dams, and the possible negative effects of climate change on hydroelectrical output given less usable precipitation and greater levels of evaporation. In addition, the current stock of nuclear power stations in Europe, Japan, and North America are aging without replacements being built. Japan and Germany have already decided to stop using nuclear power, and the retirement of the aging plants in other countries will reduce nuclear electricity output there as well. In the USA, a number of nuclear power stations have had their useful lives extended but that process has its limits, and seems to be based as much upon the loosening of regulatory standards as on successful refurbishment. There is, however, some degree of global growth in both hydroelectric and nuclear power production, the vast majority of new nuclear plants and hydro dams being situated in the developing world. Again, the middle-income developing countries will be the winners and the richer importing countries, the losers.

Exacerbating the global energy problem is the lower net energy levels of both the new fossil fuel energy sources and most of the renewables. Thus, increases in gross output may conceal decreases in the net energy available to society as more energy is required both to find and exploit any new—and usually less energy dense—sources. Any attempt to significantly offset declining oil production through the conversion of coal or gas to synthetic oil will only add to this problem, as a large amount of the energy embedded in the coal and gas is needed for the conversion process. Any large scale conversion of cars, trains, and trucks to run on electricity will also increase fossil fuel demand given the significant amounts of fossil fuels required for vehicle replacements/conversions and the large scale infrastructural changes that would be required.

The future reality may be one of no growth, or even contraction, at the global level. The wealthier countries with their aged energy infrastructures, political opposition to the use of fossil fuels and nuclear energy, and inability to outbid the developing countries for fossil fuel supplies will shoulder a disproportionate amount of the economic sacrifices. Joining them, of course, will be the poorest of the world's nations, where the decisions may be more about life and death than whether or not to buy a new car. For a country such as the UK, which has used up its once-vast coal supplies (with the largest producer seeking voluntary liquidation in 2013 [42]) and sold off its North Sea oil and gas at bargain basement prices in the 1990s, the impacts may be both difficult and tragic. The drive by many European countries to expand renewable energy sources also makes a lot of sense in an era of falling available energy exports. The intermittency of wind and solar may limit them to 20 % of electricity production at the most without expensive new smart electricity grids and advances in energy storage technologies [43]. Further, since the output of wind and solar energy is limited by external factors, fossil fuel plants will still be needed to provide extra electricity for when the sun is not shining, the clouds are thick, and the winds are absent. Even with Germany's huge focus on wind and solar, it is building 11 GW of new coal-fired power plants with the environment minister stating that such plants will be needed for "decades to come" [11].

With the need to combat climate change, the future may also be constrained by limitations on the amount of fossil fuel energy that can be safely used, as well as the amount of fossil fuel available. Instead a long and complex journey to a simpler way of living may be what stands before Earth's population, especially in rich, industrialized countries. With the need to build out a whole new energy infrastructure in the face of falling societal levels of net energy, much of the current consumer economy may have to be curtailed to free up the required energy resources. Given the fragility of modern economies when faced with contraction, as shown during the 2008/2009 financial crisis, a lack of economic growth, or even contraction, may produce a new crisis and severely handicap the society's ability to respond to the energy crisis.

References

1. Hall C (2012) Energy and the wealth of nations: understanding the biophysical economy. Springer, New York
2. Andrews S (2013) Whither shale oil? Interview with David Hughes, Post Carbon Institute. http://www.postcarbon.org/article/1816957-whither-shale-oil-interview-with-david-hughes. Accessed 5 June 2013
3. Brown J (2013) The export capacity index, Association for Peak Oil U.S.A. http://peak-oil.org/2013/09/the-export-capacity-index/. Accessed 5 June 2013
4. Berman AE (2012) After the Gold Rush: a perspective on future U.S. natural gas supply and price, the oil drum. http://www.theoildrum.com/node/8914. Accessed 5 June 2013
5. Cueille J-P (2012) The oil and gas producing countries of North Africa and the Middle East, Energies nouvelles. http://www.google.ca/url?sa=t&rct=j&q=&esrc=s&source=web&cd=9&ved=0CF8QFjAI&url=http%3A%2F%2Fwww.ifpenergiesnouvelles.com%2Fcontent%2Fdownload%2F71815%2F1530638%2Fversion%2F3%2Ffile%2FPanorama2012_05-VA_Oil-and-gas-North-Africa-Middle-East.pdf&ei=Dd4yUcvalqaZyAHbm4CYAQ&usg=AFQjCNGkWEE1EWklpn83m9xMySBglf7O2w & bvm=bv.43148975,d.aWc. Accessed 5 June 2013
6. Pirani S (2012) Central Asian and Caspian gas production and the constraints on export, the Oxford Institute for Energy Studies. http://www.oxfordenergy.org/wpcms/wp-content/uploads/2012/12/NG_69.pdf. Accessed 6 June 2013
7. Tollefson J (2013) Air sampling reveals high emissions from gas fields, Nature. 482(7384). http://www.nature.com/news/air-sampling-reveals-high-emissions-from-gas-field-1.9982. Accessed 5 June 2013
8. Zhang Y (2012) LNG set to challenge diesel's prowess, Financial Times, March 19th, 2013
9. N/A (2011) Asia leads growth in global coal production since 1980, Energy Information Agency. http://www.eia.gov/todayinenergy/detail.cfm?id=4210. Accessed 8 June 2013
10. N/A (2006) Coal production in the United States, Energy Information Administration. http://www.eia.gov/cneaf/coal/page/coal_production_review.pdf. Accessed 8 June 2013
11. Heinberg R (2009) Blackout. New Society Publishers, Gabriola Island
12. Sington D (2005) Global dimming, British Broadcasting Company. http://www.youtube.com/watch?v=nmywf7a9OlI. Accessed 8 June 2013
13. Romano F, Ramirez J, Brown T (2012) Vulnerability of U.S. water supply to shortage: a technical document supporting the Forest Service 2010 RPA Assessment, Rocky Mountain Research Station. http://www.treesearch.fs.fed.us/pubs/42363. Accessed 9 June 2013
14. Steinhurst W, Knight P, Schultz M (2012) Hydropower greenhouse gas emissions, Synapse Energy Economics Inc. http://www.clf.org/wp-content/uploads/2012/02/Hydropower-GHG-Emissions-Feb.-14-2012.pdf. Accessed 9 June 2013
15. Zittel W, Schindle J (2006) Uranium resources and nuclear energy, Energy Watch Group. http://www.energywatchgroup.org/fileadmin/global/pdf/EWG_Report_Uranium_3-12-2006ms.pdf. Accessed 10 June 2013
16. Funk J (2011) Nuclear plant surrounded by flooding is safe, Associated Press. http://www.nbc-news.com/id/43443913/ns/us_news-environment/#.UUe1IVcql8E. Accessed 10 June 2013
17. Rascoe A (2012) U.S. approves first nuclear plant in a generation, Reuters. http://www.reuters.com/article/2012/02/09/us-usa-nuclear-nrc-idUSTRE8182J720120209. Accessed 11 June 2013
18. Smith R (2012) New nuclear plant hits some snags. Wall Street Journal. http://online.wsj.com/article/SB10001424127887324731304578193880676864240.html. Accessed 6 July 2013
19. N/A (2012) Plans for new reactors worldwide, World Nuclear Association. http://www.world-nuclear.org/info/Current-and-Future-Generation/Plans-For-New-Reactors-World-wide/#.UUexb1cql8E. Accessed 20 July 2013
20. Koven P (2012) No surprises in Cameco's Cigar Lake report, Financial Post. http://business.financialpost.com/2012/02/27/no-surprises-in-camecos-cigar-lake-report/. Accessed 20 July 2013

21. van de Zwaam B et al (2003) The economics of reprocessing vs. direct disposal of spent nuclear fuel, Harvard University. http://belfercenter.ksg.harvard.edu/publication/2089/economics_of_reprocessing_vs_direct_disposal_of_spent_nuclear_fuel.html. Accessed 20 July 2013
22. Schneider M, Marignac Y (2008) Spent nuclear fuel reprocessing in France, International Panel on Fissile Materials. http://www.google.ca/url?sa=t&rct=j&q=&esrc=s&source=web&cd=3&sqi=2&ved=0CD4QFjAC&url=http%3A%2F%2Ffissilematerials.org%2Flibrary%2Frr04.pdf&ei=KZhKUfKDCseOrQGxl4GoBg&usg=AFQjCNE4exCjHapfnc3KLJc8RlLagyEz1Q&bvm=bv.44158598,d.b2I. Accessed 20 July 2013
23. N/A (2012) Processing of used nuclear fuel, World Nuclear Association. http://www.world-nuclear.org/info/Nuclear-Fuel-Cycle/Fuel-Recycling/Processing-of-Used-Nuclear-Fuel/#.UUqdCTf7R_Y. Accessed 20 July 2013
24. N/A (2013) Trial operation of rokkashi furnace, World Nuclear News. http://www.world-nuclear-news.org/WR-Trial_operation_of_Rokkasho_furnace-2101135.html. Accessed 20 July 2013
25. N/A (2012) BP Energy Outlook 2030, British Petroleum. http://www.google.ca/url?sa=t&rct=j&q=&esrc=s&source=web&cd=6&ved=0CEkQFjAF&url=http%3A%2F%2Fwww.bp.com%2Fliveassets%2Fbp_internet%2Fglobalbp%2FSTAGING%2Fglobal_assets%2Fdownloads%2FO%2F2012_2030_energy_outlook_booklet.pdf&ei=qJVQUZilKsTYyQHTrYHwDw&usg=AFQjCNGin0zrDw98JHpENanHakkl8Ba8aQ&bvm=bv.44158598,d.aWc. Accessed 20 July 2013
26. N/A (2012) Global Wind Energy Outlook 2012, Global Wind Energy Council and Greenpeace
27. Kennedy B (2012) Corn shortage hurts US ethanol production, MSN Money. http://money.msn.com/now/post.aspx?post=f28314a2-c3d3-4dfc-8a73-69b2ff6f3925. Accessed 23 July 2013
28. Giampietro M et al (2009) The biofuel delusion. The Fallacy of Large Scale Agro-Biofuels Production, Routledge
29. Angelo C (2012) Growth of ethanol fuel stalls in Brazil, Nature. http://www.nature.com/news/growth-of-ethanol-fuel-stalls-in-brazil-1.11900. Accessed 20 July 2013
30. Hall C, Dale B, Pimentel D (2011) Seeking to understand the reasons for different energy return on investment (EROI) estimates for biofuels, sustainability 2011, 3. http://www.google.ca/url?sa=t&rct=j&q=&esrc=s&source=web&cd=9&ved=0CG0QFjAI&url=http%3A%2F%2Fwww.mdpi.com%2F2071-1050%2F3%2F12%2F2413%2Fpdf&ei=670vUYLGDuSMyAHOzoD4AQ&usg=AFQjCNEmKNyJNNb8PdlA2_isXKlJQAtvUg&bvm=bv.43148975,d.aWc. Accessed 20 July 2013
31. Smil, V (2010) Energy myths and realities, AEI Press
32. Inhoff M et al (2010) Satellite supported estimates of human rate of NPP carbon use on land: challenges ahead, NASA Goddard Space Flight Center and Ames Research Center. http://www.google.ca/url?sa=t&rct=j&q=&esrc=s&source=web&cd=3&ved=0CDsQFjAC&url=http%3A%2F%2Fwww.nasa.gov%2Fpdf%2F505659main_NPP_pressbriefing_slides_MLI.pdf&ei=tMJZUcy0IKraygHD7YHwAQ&usg=AFQjCNFXlM1xk7vT2PTH2lVqwosYm5p1XA&sig2=OlGqJJl5WYe5A9tUw009BA&bvm=bv.44442042,d.aWc. Accessed 20 July 2013
33. N/A (2012) World Energy Outlook 2012, International Energy Agency. http://www.worldenergyoutlook.org/media/weowebsite/2012/WEO2012_Renewables.pdf. Accessed 20 July 2013
34. N/A (2010) 2010 Solar Technologies Market Report, U.S. Department of Energy. http://www.nrel.gov/docs/fy12osti/51847.pdf. Accessed 20 July 2013
35. N/A (2012) UK wave energy resource, Carbon Trust. http://www.carbontrust.com/media/202649/ctc816-uk-wave-energy-resource.pdf. Accessed 20 July 2013
36. N/A (2012) UK Energy Brief in 2012, United Kingdom Statistics Authority. https://www.gov.uk/government/uploads/system/uploads/attachment_data/file/65898/5942-uk-energy-in-brief-2012.pdf. Accessed 20 July 2013

37. Rembrandt K (2012) World Energy Consumption – Beyond 500 Exajoules, The Oil Drum. http://www.theoildrum.com/node/8936. Accessed 22 July 2013
38. N/A (2013) BP Energy Outlook 2030, British Petroleum. http://www.google.ca/url?sa=t&rct=j&q=&esrc=s&source=web&cd=7&ved=0CFQQFjAG&url=http%3A%2F%2Fwww.bp.com%2Fliveassets%2Fbp_internet%2Fglobalbp%2Fglobalbp_uk_english%2Freports_and_publications%2Fstatistical_energy_review_2011%2FSTAGING%2Flocal_assets%2Fpdf%2FBP_World_Energy_Outlook_booklet_2013.pdf&ei=2ndcUZ_gCsLZyQG4ioGoDg&usg=AFQjCNFYD1vJfA55lpbQL5ZNYKbs1X6tXg&sig2=7xteeYlnNRUNumTP58faYA&bvm=bv.44697112,d.aWc. Accessed 20 July 2013
39. N/A (2012) Manhattan's methane levels show natural gas as damaging to climate as other fossil fuels, EcoWatch. http://ecowatch.com/2013/04/01/manhattans-methane-levels-show-natural-gas-as-damaging-to-climate-as-other-fossil-fuels/. Accessed 22 June 2013
40. Cobb K (2012) Global oil exports in decline since 2006: what will importing nations do? Resource Insights. http://resourceinsights.blogspot.com/2012/09/global-oil-exports-in-decline-since.html. Accessed 23 June 2013
41. Rascoe A (2013), U.S. oil demand in 2012 was lowest in 16 years – IEA, Reuters. http://www.reuters.com/article/2013/02/27/usa-eia-monthly-idUSL1N0BRBG620130227. Accessed 23 June 2013
42. Rankin J (2013) UK's largest coal producer 'seeks voluntary liquidation', The Guardian. http://www.guardian.co.uk/business/2013/may/01/uk-largest-coal-producer-voluntary-liquidation. Accessed 23 June 2013
43. Palmer G (2013) Household solar photovoltaics: supplier of marginal abatement, or primary source of primary source of low emission power? Sustainability 5(4). http://www.mdpi.com/2071-1050/5/4/1406
44. Schiermeier Q (2013) Renewable power: Germany's energy gamble, Nature. http://www.nature.com/news/renewable-power-germany-s-energy-gamble-1.12755. Accessed 22 June 2013

Chapter 4
A Financial System Addicted to Exponential Growth

4.1 Introduction

4.1.1 Limitations on Energy Availability

Fossil fuels (coal, oil and natural gas) provide about 87 % of the energy utilized by humanity [1], and modern societies are completely dependent upon this massive amount of energy to maintain and grow their size and complexity. Among the fossil fuels, oil takes a special place, not only providing about 33 % of global energy [1], but also being the predominant transport fuel given its energy density and liquid form at room temperature. Since the exploration, extraction, and transportation of the other fossil fuels are themselves heavily dependent upon the availability of oil, a shortage of oil could easily lead to constraints on the supply of all the other fossil fuels. Coal, gas, and even plants (such as corn and sugarcane) can be used to produce something like oil but much of the energy is used up in the conversion process. In the foreseeable future, these processes will not be able to significantly offset declines in oil production as there are severe limitations on the rate at which the required facilities could be put in place and the scale of production that they could support. Some transport sectors could be converted to use electricity, but the huge infrastructure and vehicle changes needed would require significant amounts of the declining net energy supplies available. History demonstrates that large scale energy use transformations have taken many decades to complete [2]. Thus, if global oil production falters, or even falls, within the next decade, economic growth will almost certainly grind to a halt.

As the older, cheaper-to-produce oil fields are being depleted, their production is being replaced by much more expensive newer fields that have lower production flow rates, thus compounding the problem. Each new oil resource tends to produce less net energy per barrel of oil, and fewer barrels of oil, when compared to earlier resources. The energy return on investment (EROI) for conventional oil discoveries in the USA has fallen from about 100:1 in the mid-twentieth century, to 18:1 a few years ago, and as low as 10:1 for the latest fields [3]. For the unconventional oil deposits, such as deep sea oil, tar sands and shale oil, the EROI has been

R. Boyd, *Energy and the Financial System,* SpringerBriefs in Energy,
DOI 10.1007/978-3-319-04238-1_4, © Roger Boyd 2013

estimated in the range of 5:1 [3]. Even with apparently steady oil production levels since 2005, the net energy being provided to society has fallen. Almost certainly, it will soon become impossible for the newer oil fields to offset the flow declines in the older fields. Global production will fall, and the decline in net energy will accelerate. With that, the global economy will start to contract, with the heavily oil-dependent global supply chains being one of the first casualties. In fact, there is considerable evidence that this process has already started.

Natural gas production provides about 24 % of global energy usage [1]. The outlook is for at least another decade of increasing natural gas production, but with probable reductions in EROI as more remote and difficult-to-produce deposits are utilized. The increasing transportation of liquefied natural gas (LNG) by ship also results in a reduction of the net energy available as it consumes more than 10 % of the energy embodied in the gas being transported [4]. Coal production, which provides the balance of the fossil fuel supply, has increased rapidly in the past decade. The outlook is for continuing increases in coal production, but a likely reduction of net energy as lesser grades, and deeper deposits of coal, are used. As touched on above, the peak production of coal and gas may arrive sooner as oil supplies dwindle and all stages of exploration, extraction and transportation become more and more expensive.

The alternatives to the fossil fuels currently provide only 13 % of global energy supplies [1], with the vast majority coming from hydro-and nuclear power. The former provides high net energy output, but its growth is limited by available sites, especially in the advanced industrial countries. The latter has an EROI of no greater than 5:1 [3], and thus provides no real replacement for high net energy supplies. The "new renewables" such as wind and solar energies provide only 2 % [1] of current global energy supplies, and even with continued rapid growth, would not significantly offset the decline in fossil fuels, let alone be able to power continued growth. In the end, these alternatives, too, are heavily dependent on oil supplies from the extraction of raw materials, through the production of required parts, to installation and maintenance.

The overall picture for the global economy is one of, at best, severely constrained energy supplies, with greater amounts of the available energy being diverted back into the production of that energy rather than being available for other activities. There are those arguing that greater energy efficiency in some countries has allowed them to reduce energy demand, thus freeing up energy for other uses even though production is flat. While this may be the case, an equal, somewhat countervailing argument is that this reduced energy usage is a result of high prices and faltering economies causing people to use less energy than they would otherwise. There is also the possible impact of abrupt climate change [5], which may focus society on seriously reducing greenhouse gas emissions well below the current levels [6]. Such policies would lead to strict limitations on fossil fuel use, producing the same effects as if the fossil fuel supplies had been depleted. With economic growth dependent upon increasing energy supplies, this cutback will lead to economic contraction and stagnation. What will that mean for a financial system that has evolved in the age of cheap energy supplies provided by fossil fuels?

4.1.2 *The Financial System as a Time Machine*

The financial system acts like a time machine, bringing our expectations of the future into the present. In the period leading up to the 2008 financial crisis, the vast majority of investors were too optimistic about the future. Property prices would keep rising; the world economy, turbo-charged by globalization, would keep growing; only blue skies lay ahead. Increasing share prices, and the growth in lending reflected these positive expectations. Eventually the reality of mounting mortgage defaults chipped away at the positive mood, and was reinforced by the failures of some over-exposed funds and institutions like Bear Sterns. In late 2008, the failure of Lehman Brothers delivered the final nail in the coffin to any remaining optimism and darkness overwhelmed the financial system. Share prices crashed, lending fell, and financial companies previously thought to be safe got into trouble. Suddenly, no one knew what or whom to trust. Without the lifeblood of confidence, the financial system ground to a halt. Only unprecedented intervention by many governments—and the infusion of trillions of dollars, euro, yen, pounds, and other currencies conjured up by central banks—saved the day and pulled the world back from the brink of a new Great Depression.

The current financial system relies upon the ongoing payment of interest on loans, which in turn depends on increases in company profits and individual incomes. Continuing economic growth is required to provide those increasing profits and incomes to pay that interest. Without this growth, the financial system as currently structured would cease to function. Everything from personal savings to pension funds, insurance company reserves, and bank reserves are invested in the financial markets; their malfunctioning would cause severe economic and social distress. For example, as shown in Fig. 4.1 below, US household assets total US$ 78.2 trillion, which is nearly five times the yearly output of the US economy [7]. Of this US$ 78.2 trillion, only US$ 24.6 trillion are tangible assets such as real estate and durable goods. All of the rest are financial assets whose value is dependent upon the outlook for future growth [8]. And as we saw in 2007 and 2008, real estate prices are also heavily dependent upon the prospects of future growth. Lack of growth would also put a large slice of the tangible US$ 24.6 trillion in assets at risk.

The level of household assets held within the financial system does vary with richer countries tending to have a much higher share of financial assets, although there can also be significant differences between countries with relatively equal levels of income. Within financial assets, the mix between shares and other assets (such as deposits and bonds) also varies between countries. The higher the share of financial assets is for a given country, especially equities, the higher the linkage of wealth to future growth assumptions (Fig 4.2).

The larger the amount of financial assets when compared to the size of a country's economy, the greater impact a lack of growth will have. It is the underlying economy which provides the ability to pay the interest and principal on debts and to provide the profits underpinning share prices. The richer countries tend to have much greater levels of financial assets relative to the size of their economies. In addition, the level of debt has multiplied in the mature industrialized countries in the

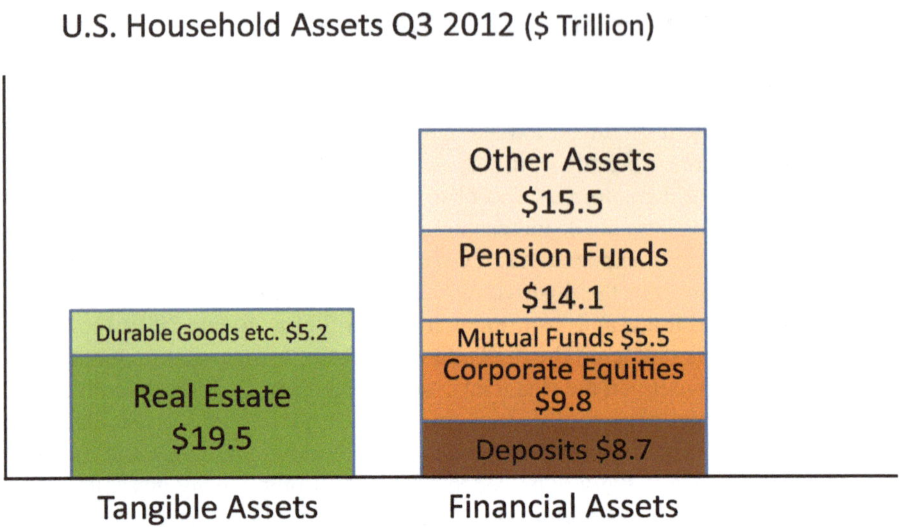

Fig. 4.1 U.S. Household Assets (Source: US Federal Reserve Z1)

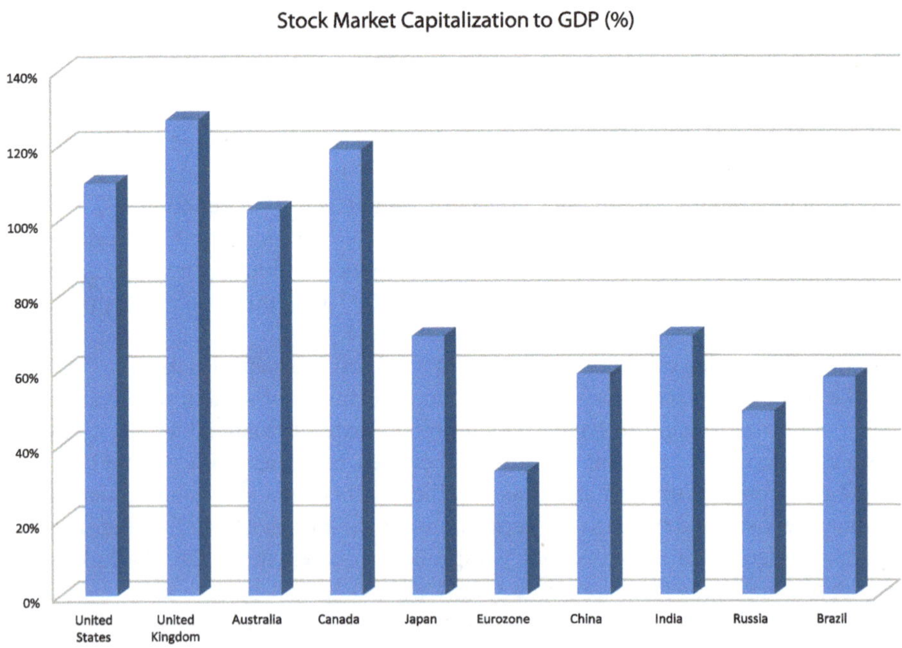

Fig. 4.2 Stock Market Capitalization to GDP for Selected Countries (Source: Federal Reserve Board of Saint Louis)

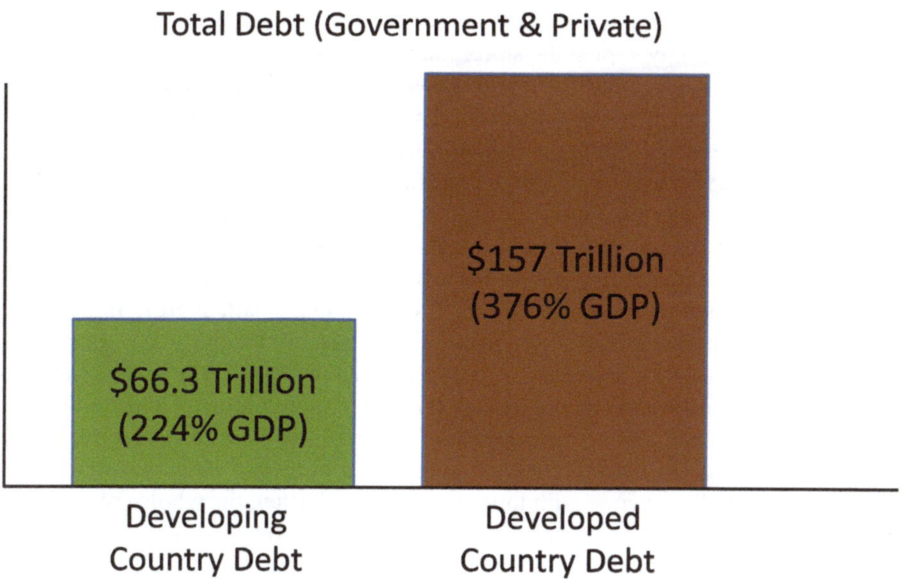

Fig. 4.3 Total Debt for Developing and Developed Countries (Source: ING)

past two decades, providing a much large booby-trap to trigger if growth comes to a halt. Thus, wealth, or at least the illusion of wealth, within some of the richer countries is dependent to a much greater extent on assumptions of continued growth, than in the mid- and lower income. As a lien on future energy usage, debt increases the ability of an individual to access part of the available energy supply to generate goods and services which hopefully will allow that individual to at least meet the principal and interest payments on that debt. For the lender, the interest payments provide them with greater access to the available energy.

The level of debt has grown at a much slower pace in the mid-income developing countries, such as Brazil, China, India, and Russia. Thus the size of the impact from a debt-deflation cycle triggered by the realization of physical growth constraints will be much less in these countries (Fig. 4.3).

The financial system can be said to have the same bounded resilience that many natural systems have. It can function within a certain range of variability, but anything outside that range (such as the 2008 crisis), brings it to a grinding halt. Exacerbating this problem, the global economy as a whole has also become much more brittle and less resilient to shocks through the drive for greater integration and efficiency. Such things as local independent financial institutions, local supplies of goods, and even locally grown food have been removed in the name of efficiency and globalization. Without them, there are fewer factors of resistance to stop crises rippling throughout the economy. Instead, shocks and crises are very quickly and efficiently transmitted around the globe. In 2000, as I drove down a British motorway looking for gas during the blockade of oil depots by angry haulage drivers, there was a general worry that food supplies would run out. With only three days worth of food generally kept at the stores, and fresh stocks coming from centralized depots supplied from around the world, the lack of fuel oil for the delivery trucks, ships, and planes threatened a general food shortage. Only swift government intervention averted a crisis. Some years later, floods in Thailand disrupted electronics manufacturing around the world, as a significant amount of the global production of certain components had been consolidated into that country to take advantage of lower costs. It is as if we had removed one of our kidneys, in the same way that we consolidate manufacturing plants, because having two was seen as needless and expensive duplication. Such a move may provide some small, short term benefits, but the loss of the remaining kidney would then cause our death.

Duplication and redundancy can seem inefficient but may turn out to be lifesavers. Dwindling net energy and climate change threaten ongoing growth, and are making the environment that human societies exist within much more unstable. Yet at the same time, through increased levels of global integration and efficiency, humanity is reducing its ability to cope with those challenges. Further, the whole enterprise is completely dependent upon a financial system that cannot function without relatively stable conditions and an expectation of uninterrupted growth. In the not-too-distant future, due to either the peaking of net energy supplies or climate change-related limitations upon fossil fuel usage, humanity will likely face the end of long held growth expectations. Instead, what appeared to be a reliable path of growth will be shown to have been an aberration in human existence, driven by the amazing gifts of those fossil fuels. The fragile, growth-addicted financial and economic system will then become a huge series of booby-traps, greatly exacerbating the resulting crisis and constraining society's ability to take necessary corrective actions.

4.2 Share Prices

A company's market capitalization (the number of outstanding shares multiplied by the current share price) is generally a product of its current profits and the perceived opportunity for future growth in those profits. Profits are what are left from revenues after paying the operating costs of the business (including taxes and depreciation) and any debt payments. As costs tend to be more fixed than revenues, any change in revenues will have a much bigger impact on profits, positive or negative, if everything else stays the same. Growing profits depend upon the company's relative competitiveness, the growth potential of its segment of the economy, and overall economic growth. For example, currently Google has a high share price relative to its profits (referred to generally as the "price/earnings multiple") as it does business in a high growth segment of the economy, and is very competitive in that segment. Yahoo has a much lower relative share price, as it is much less competitive. General Electric has a much lower share price multiple than Google. Although it is very competitive in the segments of the economy in which it does business, they tend to have lower relative growth.

If a future of no growth, or even contraction, becomes more likely, then it will be seen that even highly competitive companies in promising market segments will have a harder time growing revenues and lesser companies will be unable to stem a fall in revenues. This will greatly reduce the expectation of future growth in profits, which will cause share price to earnings multiples to be significantly reduced. The resulting downward movement in share prices will be of the scale usually called a "crash" from which there will be no recovery. The majority of stock price values, which reflect the mirage of future assumed growth, will be removed. With those reduced values, a very large slice of the world's financial assets will be removed.

4.3 Bond Prices

Bonds are a form of loan where pieces of the loan can be bought and sold as financial securities. The issuer sells the bonds at say $ 100, and agrees to pay fixed interest payments, say $ 10 per year, until the bond matures after an agreed period when the $ 100 is paid back. In this case, the bond is said to pay 10 % interest (since $ 10 is 10 % of $ 100). The fixed nature of the interest payments is good for the issuer, as it pre-determines what the future cash flows will be for this bond.

The first buyer of the bond may not want to hold it until maturity, and would therefore need to find others to sell it to. Conditions however, may have changed since the original purchase, and these changes would affect the interest rate that potential buyers would want before purchasing the bond. If, for example, general interest rates have been lowered by the government central bank, then an interest rate lower than the original 10 % on the bond might still be attractive. On the other hand, if potential buyers thought that the issuer's risk of defaulting had increased, they would demand a higher interest rate to offset that risk.

The interest payments on the bond are fixed though, so how can a different interest rate be paid? This is solved through varying the price of the bond, which changes the percentage that the fixed interest payment represents. For example, if the seller can find a buyer at an interest rate of 5 %, then the bond bought at $ 100 with a $ 10 fixed annual interest payment, could double in price to $ 200; $ 10 (fixed interest payment) is 5 % of $ 200. If, on the other hand, the seller could only find buyers at an interest rate of 20 %, then the bond would need to halve in price to $ 50; $ 10 (fixed interest payment) is 20 % of $ 50.

A general realization that the future will be one of no growth, or even contraction, would greatly affect the perception of the issuer's ability to pay back the interest and principal. This will be for both private companies whose revenues will be lower than previously thought, and governments whose tax revenues will be less than previously planned for. These reduced revenues and will greatly increase the risk of nonpayment by the issuers, and thus any new buyers will demand higher levels of interest to compensate for the higher risk. Bonds generally will then fall in price to reflect the higher required interest rates. This can become a self-feeding process where the higher interest rates required make the issuer riskier (as new borrowings will be at the higher interest rate), which in turn causes buyers to ask for even higher interest rates.

In cases where the required interest rates become too high for the issuer to pay on new borrowings a debt restructuring, or even a full default will be the result. In a debt restructuring the value of the outstanding bonds is cut to make the issuer less risky by reducing the level of its outstanding debt. This is the unfortunate place that banks in Cyprus found themselves in, at the start of 2013, as the Greek debt restructuring greatly reduced the value of their large holdings of Greek government bonds. In a full default, the owner of the bonds loses everything.

General uncertainty about the future may also deter investors from buying bonds from companies and governments. An increasing number of issuers would exit the bond market completely. With banks also probably unwilling to lend to such borrowers (see banks section below), cash flow problems, default, and bankruptcy could rapidly affect significant numbers of borrowers. The effect would cascade through the economy as some borrowers became unable to make payments to others, such as their suppliers, thus reducing the viability of still more issuers.

4.4 Mutual Funds

Mutual funds provide an avenue for relatively small investors to have a third party buy stocks and bonds for them, while sharing the management and transaction costs across many other small investors. The global mutual fund industry has $ 26.8 trillion in assets, with nearly half of that being held by US mutual funds. Equity holdings comprised 45 % of the US funds, with bonds at 26 %, and money market funds (holding short-term bonds) at 21 % [9]. In effect, at least 71 % of these mutual fund assets are heavily dependent upon future growth assumptions. Financial planners support such a high exposure, as they assume that growth will continue [10]. However, as share and bond prices decline very significantly, reflecting a future of no growth or contraction, mutual fund values will also fall significantly. This will be a rerun of the 2008–2009 decline although, probably larger, and with no real recovery.

4.5 Pension Funds and Annuities

Pension funds take contributions from individuals or employers and invest them in stocks and bonds (generally) seeking an assumed rate of return to provide for future pension payments. These can be of a "defined benefits" type paying out a regular monetary amount which can be related to an individual's earning prior to retirement, or of a "defined contributions" type where payments are completely dependent upon contributions and investment returns. There are also annuities where invested funds are used for a guaranteed regular income but any remaining funds on death go back to the company. In 2012, global pension fund assets stood at $ 29.7 trillion, with the USA holding more than half of that amount, and Japan, the UK, Australia, Canada, and the Netherlands each having over $ 1 trillion in pension fund assets [11]. As with mutual funds, stocks make up a significant percentage of assets. For example, the Canadian Pension Plan Investment Board holds 50 % of assets as shares, and 33.1 % as bonds [12], which is close to the average mix of retirement funds globally [11]. Such an asset mix provides a very large exposure to reduced equity and bond prices, together with lower future rates of return.

Even without a lack of future growth, the future return assumptions of the vast majority of pension funds are not realistic, and are thus overstating the ability to pay future pension amounts. Only in 2012 did the California Public Employees' Retirement System reduce its assumed rate of return from 7.75 to 7.5 % [13], but the new rate is still well above a realistic value. Other pension funds are just as overoptimistic such as the Illinois Teachers Retirement System at 8 % [14], and the Canadian Pension Plan assuming 4 % on top of inflation [15]. Such returns may have been gained in previous years, but are now unrealistic given the slower growth of debt-laden economies and the very low interest rates required to support the servicing of that debt. The dependency on financial returns to fund future payouts is shown by the Ontario Teachers' Pension Plan, where 77 % of the growth in assets in the past decade was due to investment returns, against only 23 % for pension contributions [16]. As the value of the pensions holdings of stocks and bonds collapse in price, reflecting a future of no growth or contraction, the future payouts that those investments will support will be greatly reduced. Reductions in future rate of return assumptions, from ones too optimistic for even future growth scenarios to ones reflecting a no-growth world, will very significantly reduce those payouts further.

For company-run defined-benefits pensions, the company could inject more money to rebuild the investment portfolio. With an outlook of constrained or falling profits few companies, if any, will be ready or able to do that. For annuities the providing financial organization could make up for the investment losses to meet the guaranteed payouts, but they will be under severe profitability stress as well, making them unwilling or unable to contribute the required amounts. Governments will also be faced with lower forecasts for taxes; stopping them from rescuing failed private pension schemes. Thus, for defined benefits pension plans and annuities, the expected payments will not happen, either through negotiated reductions or default. For defined contributions plans, the lower asset values will automatically flow into lower payouts. Reductions in taxes due to no growth or contraction, together with state pension scheme investment losses, will also lead to lower pensions for both state employees and for those drawing universal state pensions. Overall, current and future pensioners will see very significant falls in their current and planned incomes. In what will seem as a bolt from the blue, the realization of a future of no growth or even contraction will greatly reduce the living standards for current and future retirees. Those that have enjoyed many years of good pensions prior to the general acceptance of constraints on future growth, will come to be seen as the lucky ones who enjoyed the fruits of the fossil fuel era to the full before time ran out.

4.6 Insurance Companies

It is not generally understood that an insurance company is very much like a pension fund. It charges insurance premiums and invests them predominantly in shares and bonds to pay for future insurance claims. The weighting is generally much more towards bonds than it is for pension funds. In the US insurance industry, 70 % of investment assets are held in bonds, with only 10 % in stocks [17]. The industry employs statisticians (or actuaries) to calculate probable future claim amounts, in much the same way that pension funds calculate future pension payments. The returns on an insurers' portfolio of stocks and bonds are also calculated using predictions of the future. These predicted returns must be enough to pay for the predicted future claims. Otherwise, it risks being declared insolvent. Consequently, the insurance company would need to raise insurance premiums significantly, greatly restrict what it will insure, and reduce its payout on claims. The acceptance of a much less optimistic economic future will collapse both bond and share prices, and lead to predictions of lower future rates of return. As a result, the insurance company may be in breach of regulatory capital guidelines designed to make sure that it can meet future claims, and will need regulatory forbearance until it can rebuild its reserves.

At the same time, climate change-related insurance claims might be increasing which, along with higher claims, could result in a shortage of financial reserves. For areas where predictions of future climate change show increasing losses, for example in hurricane- or tornado-prone areas and low-lying flood plain properties, insurance premiums will also rise to reflect these increased risks. Higher and higher premiums will be the result. Some areas may be declared as uninsurable given that the high probability of events that will render insuring them unprofitable if not impossible.

The net result is that large segments of current insurable properties, activities, and events may cease to be insurable due to greatly increased premiums or a complete lack of available insurance. Some uninsurable properties will fall in price significantly, and some investments stopped or redirected, due to the lack of available insurance. Waterfront property values in places such as the Florida Keys, Miami, or even low-lying areas such as London, New York, and Shanghai, may collapse decades before the actuality of any flooding. The mere high future probability of such flooding may render them uninsurable and possibly unsellable. As noted earlier, the financial system acts like a time machine, bringing our expectations of the future into the present; the operation of the insurance market will pull the possible effects of these future events into the present.

4.7 Bank Lending

When banks are able to receive regular repayment of most of their loans, banking is one of the most profitable legal businesses in the world, with perhaps only a gambling license offering better returns. In modern banking, money is created out of thin air. When a loan is given the money appears as a debt to the borrower and as an asset to the bank on the bank's balance sheet. When a loan is paid back, the money loaned zeros out on the bank's books, but the bank gets to keep the interest payments. Charging for the use of something created from nothing! This power of money creation should be the property of the citizenry, but over time the banks have largely managed to gain control of the power to create money and the rate of interest charged.

This money creation provides the financial oil that greases the modern economy, and without it the economy would grind to a halt. In fact, this debt-based money is a lien on energy, with the assumption that if more money is created, more energy will be brought out of the ground to do the work that gives meaning to that money [3]. Otherwise inflation will accrue, as it is far easier to "print" money than mine new energy. When the effects of peak oil and failing sustainability become visible, a future of no growth or contraction will enter the public's consciousness, and the assumptions under which loans are made will have to be revised. At first, there will probably be a general reduction, or even halt, to new lending as banks try to understand the implications of the new future. Lending will eventually restart, but with much tougher criteria given the lack of future growth to cover the interest payments. Gone will be the internet loans, and back will come the tough discussions with the local bank manager. The vast majority of bank lending is against property values (mortgages, for example), and declining or stalled growth will reduce those values. The resulting reduction in the available collateral to back such lending will then act to constrain lending. This will reduce property values that historically have risen steadily over the long term. The lower values will reduce lending further in another vicious circle. Existing borrowers may then increasingly default on, or even walk away from their loans, creating increasing credit losses for the banks.

Banks that reappraise their current loans to reflect the new future reality due to higher risks of defaults will also risk insolvency, as the value of those loans will be significantly less, wiping out the bank's capital. It will not help that much of that capital will have been invested in devalued bonds, as the banks in Cyprus learned in 2013 [18]. Most probably, the regulators will allow the banks to misrepresent their loan values, as has happened with US banks after the housing property crash (since 2007), to try to forestall a general banking crash [19]. With billions and trillions of dollars worth of derivatives—although only leveraged bets on share and bond prices—also on banks' books, such efforts will probably be in vain. With the collapse of the banking system, the modern financial system that drives and coordinates economic activity will cease to function, and along with it, the real economy which is needed to help combat the effects of energy shortages and climate change. In addition, unemployment will rapidly rise as failing enterprises, and those struggling for survival, shed workers in a desperate attempt to cut costs. These now unemployed workers will cut consumption (deepening the economic impact), and default on loans, producing another vicious debt-deflation cycle for the financial sector. Once the boundaries of sustainability have been breached, the move to a new, less pleasant state can be both chaotic and scarily rapid. This is exactly what happened in the 1930's Great Depression and what came very close to happening in 2009.

4.8 Investment Decisions

When companies make investment decisions, they forecast the possible future financial returns on an investment and compare it to the amount of money to be invested. In a world where no growth, or ongoing contraction, is the norm, such future returns will be much lower than those expected in a growing economy. This will mean that many fewer investments would be likely to provide sufficient returns to pay back the investment and provide an acceptable rate of return, with renewable energy investments being among them. Investment spending therefore will be greatly reduced, further decreasing economic activity in general, and intensifying the effects of the financial crisis.

4.9 Summary

The current global financial system is extremely brittle. It is completely dependent upon continued exponential growth, and moreover, is easily broken by relatively minor issues that it rapidly magnifies and spreads. This has been shown repeatedly in the past two decades with one crisis after another, each of which had the ability to cause a global financial crash were it not for government intervention: The Mexican Peso crisis (1994) [20], Asian financial crisis (1997) [21], Russian default (1998) [22], and resultant failure of Long Term Capital Management Hedge Fund (1998) [23], bursting of the "Dot-com Bubble" (2000) [24]. The bursting of the US housing bubble in 2007/2008 [25, 26] led to a crisis of such magnitude that the usual government intervention could not stop a crash. Unprecedented levels of intervention and money printing (figuratively of course, as money is overwhelmingly created electronically these days) were required to stabilize the situation. The general economy has also been made more fragile by the drive for greater levels of efficiency and globalization. An issue in one part of the economy can now be rapidly transferred to another due to the "just-in-time" driven reduction in local supplies and the sheer complexity of modern supply chains.

When the inevitable limit to growth gains wide acceptance, a financial contraction will rapidly cascade through the economy on a global scale, rendering much of it inoperable or severely impaired. There are some signs that this may have already started in the OECD countries where growth has virtually ceased, or certainly has not reached the rates of previous decades. Unlike in 2009, simple money printing will not rectify the situation, as this crisis will be caused by external physical limits, rather than just internal financial problems. There may be short-lived and regional recoveries, but the economy will not return to the pre-crisis levels of output. The fundamental underpinnings of the current financial and wider economic arrangements will have been removed; the end of future growth will have entered the general consciousness. The dislocation this causes to everyday decisions, together with the sheer scale of the financial losses incurred, will constrain if not derail any recovery. The severe consequences may also divert attention away from actions needed to deal with climate change, and instead focus them towards short-term actions aimed to reduce unemployment and help businesses to survive. A preview of such issues has been seen in Europe where business pressure prevented climate-friendly policies, such as a higher carbon price, which were said to be bad for growth. As well as these policy issues, a severely damaged financial system will greatly reduce the funding for new fossil fuel and renewable energy investments. Together, these problems may cause estimates of future fossil fuel supplies and renewable energy growth to be seen as overly optimistic.

The severity of the impacts may vary between countries though, based upon the levels of financial assets and debts that a given country has and its ability to access the dwindling sources of fossil fuel energy. Earlier in this chapter it was noted that the mature industrialized economies have much higher percentages of their assets held in financial markets, and have much higher levels of debt, than middle-income and poor countries. The middle income countries may also be able to out-bid the mature economies for energy at the margin as they are still exploiting the lower-hanging-fruit, higher return, uses of energy that the mature countries exploited long ago. The recent reduction in oil usage in many of the mature economies, including the US, may be an omen of such a reality [27]. These reductions have allowed China and India to increase their oil consumption in the context of generally flat global oil production. The Export Land Model proposed by Jeffrey Brown [28] argues that energy exporters will utilize more and more of their energy production as their populations grow and levels of affluence increase. Even with stable energy production this will mean that energy exports will fall over time, and an increasing number of countries will ultimately cease to be energy exporters. With middle income developing countries outbidding the mature industrialized countries for energy supplies, and the exporting countries keeping greater amounts of their production for their own uses, the impacts of falling net energy availability will be most highly felt in the mature industrialized importing countries, such as Europe, Japan, and the United States. These are the very same countries that have the highest levels of financial assets as a share of net worth, and the highest levels of debt.

It has been estimated that a no-growth economy could only operate with debt levels of 40 % [29] and under, against the current levels of 300 % plus for most countries. In previous low to no-growth periods regular debt write-offs (known as jubilees) were required. Otherwise the lenders would end up enslaving the rest of society as debt compounded while economic output did not [30], most probably leading to a general revolt against the wealthy. This could be the fate that awaits much of the population without such debt forgiveness. The reduction in debt will be paralleled by a reduction in the financial industry. Without a complete redesign of the financial system, including very significant reductions in outstanding levels of debt, that system will act as a multiplier of the peak energy and climate change crises, and handicap society's ability to respond. Again, the impact will be most felt in the mature industrialized countries that import energy, such as Europe, Japan, and the USA. Further, much of the efficiency and globalization drives of the past few decades will also need to be reversed to provide society with a level of resilience able to withstand future crises while still being able to adequately function. Such changes would be incredibly challenging in the best of times, but without them, modern societies will fail in the face of energy constraints. The longer humanity waits to make such changes, the harder those changes will become, as net energy, climate change, and other challenges wield greater and greater impacts upon society.

References

1. N/A (2012) BP Statistical Review of World Energy June 2012. British Petroleum. http://www.bp.com/assets/bp_internet/globalbp/globalbp_uk_english/reports_and_publications/statistical_energy_review_2011/STAGING/local_assets/pdf/statistical_review_of_world_energy_full_report_2012.pdf. Accessed 25 July 2013
2. Smil V (2010) Energy transitions: history, requirements, prospects. Praeger, Westport
3. Hall C, Klitgaard K (2012) Energy and the wealth of nations: understanding the biophysical economy. Springer, Berlin
4. n/a (n/a) Liquefied natural gas chain, natgas.info. http://www.natgas.info/html/liquefiednaturalgaschain.html. Accessed 25 July 2013
5. McKinney K (2012) Why Arctic Sea ice shouldn't leave anyone cold. Arctic Sea ice blog. http://neven1.typepad.com/blog/2012/08/wasislac.html. Accessed 25 July 2013
6. Hansen J et al (2008) Target atmospheric CO_2: where should humanity aim? Columbia Univesrity. www.columbia.edu/~jeh1/2008/TargetCO2_20080407.pdf. Accessed 25 July 2013
7. Irwin N (2013) Huzzah! The U.S. economy is 3 percent bigger than we thought. Thanks, George Lucas!, The Washington Post. http://www.washingtonpost.com/blogs/wonkblog/wp/2013/04/22/huzzah-the-u-s-economy-is-3-percent-bigger-than-we-thought-thanks-george-lucas/. Accessed 25 July 2013
8. Durden T (2012) US household assets: $78.2 trillion, liabilities: $13.5 trillion; net worth: $64.8 trillion. Zero Hedge. http://www.zerohedge.com/news/2012-12-06/us-household-assets-782-trillion-liablilties-135-trillion-net-worth-648-trillion. Accessed 25 July 2013
9. n/a (2013), 2013 investment company fact book: recent mutual fund trends. Investment Company Institute. http://www.icifactbook.org/fb_ch2.html. Accessed 27 July 2013
10. McFeat T (2012) Many Canadian's retirement plans in dire need of a reality check. Canadian Broadcasting Corporation. http://www.cbc.ca/news/business/taxseason/story/2012/01/10/f-rrsp-retirement-reality-check.html. Accessed 27 July 2013
11. n/a (2013) Global pension assets study 2013, Towers Watson. http://www.towerswatson.com/en-ZA/Insights/IC-Types/Survey-Research-Results/2013/01/Global-Pensions-Asset-Study-2013. Accessed 27 July 2013
12. n/a (2013) Asset mix. CPP investment board. http://www.cppib.ca/Investments/Total_Portfolio_View/asset_mix.html. Accessed 28 July 2013
13. Marois M (2013) Calpers committee proposes lowering rate of return to 7.5%. Bloomberg. http://www.bloomberg.com/news/2012-03-13/calpers-committee-proposes-lowering-rate-of-return-to-7-5-1-.html. Accessed 28 July 2013
14. n/a (2012) TRS revises assumed rate of investment return. Teachers Retirement System of the State of Illinois. http://www.google.ca/url?sa=t&rct=j&q=&esrc=s&source=web&cd=18&ved=0CGwQFjAHOAo&url=http%3A%2F%2Ftrs.illinois.gov%2Fsubsections%2Fpress%2F2012%2FSept21_12.pdf&ei=nA-uUd6-J6PFyAGr6ICQAw&usg=AFQjCNH5wmuok-Wxav7_IeiafMaQQuW02A&sig2=hb7WHkUod-PT2a5qiZ0elA&bvm=bv.47244034,d.aWM. Accessed 28 July 2013
15. Falconer K (2013) CPP fund hits $183B, posts 10.1 % investment return. PreHub Canada. http://www.pehub.com/202666/cpp-fund-hits-183b-posts-10-1-investment-return/. Accessed 25 July 2013
16. Shecter B (2013) Teachers' pension returns 13 % in 2012, but continues to battle demographics. Financial Post. http://business.financialpost.com/2013/04/02/teachers-pension-returns-13-in-2012-but-continues-to-battle-demographics/. Accessed 29 July 2013

17. n/a (2011) Capital markets special report. National Association of Insurance Commissioners & the Center for Insurance Policy and Research. http://www.naic.org/capital_markets_archive/110819.htm. Accessed 29 July 2013
18. Goldman R (2013) The Cyprus crisis explained (like you're an idiot). ABC News. http://abcnews.go.com/Business/cyprus-crisis-explained-idiot/story?id=18782212#.UbDOi9jD7wA
19. Scannel K (2009) FASB eases mark-to-market rules. The Wall Street Journal. http://online.wsj.com/article/SB123867739560682309.html. Accessed 29 July 2013
20. Lustig N (1995) The Mexican Peso crisis: the foreseeable and the surprise. Brookings Institute. http://www.google.ca/url?sa=t&rct=j&q=&esrc=s&source=web&cd=5&ved=0CEoQFjAE&url=http%3A%2F%2Fwww.brookings.edu%2Fviews%2Fpapers%2Fbdp%2Fbdp114%2Fbdp114.pdf&ei=bNCwUf6RAob_yQG_6YDACw&usg=AFQjCNGav7B7M9jEucF8Lip-f9l7PMd7bA&sig2=wZDjy1ZnTD00T61vbSY8QQ&bvm=bv.47534661,d.aWc. Accessed 5 August 2013
21. King M (2001) Who triggered the Asian financial crisis? Rev Int Polit Econ 8(3):438–466 (Autumn)
22. Chiodo A, Owyang M (2002) A case study of a currency crisis: the Russian default of 1998. Federal Reserve Bank of St. Louis. http://www.google.ca/url?sa=t&rct=j&q=&esrc=s&source=web&cd=3&ved=0CDcQFjAC&url=http%3A%2F%2Fresearch.stlouisfed.org%2Fpublications%2Freview%2F02%2F11%2FChiodoOwyang.pdf&ei=X9GwUcKAPOScyQHYlYHoCw&usg=AFQjCNE5iSC91OrH-ayaU-ewGjqRoRjfoA&sig2=9zlYPJT0B6-Bw3BV47bG6A&bvm=bv.47534661,d.aWc. Accessed 5 August 2013
23. Lowenstein R (2001) When genius failed: the rise and fall of long term capital management. Random House, New York
24. Goodnight GT, Green S (2010) Rhetoric, risk, and markets: the dot-com bubble. Quarterly Journal of Speech. http://www.academia.edu/246551/Rhetoric_Risk_and_Markets_The_DotCom_Bubble. Accessed 5 August 2013
25. Duca JV, Muellbauer J, Murphy A (2012) Credit standards and the bubble in US house prices: new econometric evidence. Bank of International Settlements: BIS Papers No. 64
26. Demyanyk Y, Van Hemert O (2008) Understanding the subprime mortgage crisis. Federal Reserve Bank of St. Louis: Working Paper 2007–05
27. Tverberg G (2013) Why is US oil consumption lower? Better gasoline mileage? Our finite world. ourfiniteworld.com/2013/01/31/why-is-us-oil-consumption-lower-better-gasoline-mileage. Accessed 5 August 2013
28. Brown J (2013) Commentary: the export capacity index. Resilience. http://www.resilience.org/stories/2013-02-18/commentary-the-export-capacity-index. Accessed 5 August 2013
29. Jackson T (2009) Prosperity without growth: economics for a finite planet. Routledge, London
30. Graeber D (2011) Debt: the first 5,000 years. Melville House, New York

Chapter 5
So What Can I Do?

5.1 Introduction

Most of what we know as wealth is a claim upon future economic growth, growth that has, to date, been made possible by huge supplies of cheap energy. Now, with increasing constraints on cheap energy, the vast majority of that wealth will be shown to be a mirage. The "financialization" of the advanced industrial economies in the past few decades has exacerbated this problem by greatly increasing such "wealth" in the form of debt and equity holdings. In addition, globalization has reduced the resilience of the economy through placing efficiency well above diversity and sustainability. Thus, the impending impact of cheap energy constraints may cause a cascade of problems facilitated by the complex webs of global supply chains and financial linkages. The crisis of 2007–2009 is only the latest example of how a crisis in one financial area can be rapidly transformed into a global crisis and near collapse. We may not be able to predict when the next crisis will hit, but we can predict that it may spread extremely rapidly. Trying to take defensive actions once the crisis has started may turn out to be a relatively futile gesture. Instead, such actions need to be made beforehand. These actions would have to take into account the exposure of a person's job to the financial system as well as the exposure of their investments within that same system.

Location may play a significant part in how someone will fare once cheap energy constraints become evident. The best individual plans may come to nothing if the country, region, city, or town where one lives is greatly dependent upon either cheap energy from somewhere else or the financial system. The energy importers, especially those that are mature economies, will be the first to be hit and the hardest. Locations with a large share of employment and economic activity coming from the financial system will also be heavily hit as that system crashes. In addition, countries which are dependent upon exchanging financial I.O.U.'s (debts) for critical imports may find that the supporting financial arrangements—and the requisite trust—are suddenly in very short supply.

Some may believe that the government, or "the powerful," will wake up and skillfully navigate society through the crises. History does not provide much evidence of elites grasping the reality of a given situation and taking corrective action.

R. Boyd, *Energy and the Financial System*, SpringerBriefs in Energy,
DOI 10.1007/978-3-319-04238-1_5, © Roger Boyd 2013

Instead it is full of societies driving themselves over the edge while stubbornly refusing to accept reality. Psychology and sociology both support a relatively pessimistic view of the ability for individuals and societies to accept fundamental challenges to their worldview and to take the painful actions necessary to mitigate the consequences of continuing on the same path. The advanced industrial societies have spent 200 years creating and exporting to the rest of the world an edifice of beliefs based upon perpetual growth. The elites of most nations, those who benefit the most from these beliefs, have been thoroughly indoctrinated and are, therefore, the most handicapped in accepting a new reality. One example of such willful blind belief is the resistance of mainstream economics to the reality of ecological and geological limits to growth. Resistance can be overcome, as was the case with the scientific revolution with respect to church dogma, but this took many decades, if not centuries. We do not have decades, and certainly not centuries. Since some countries will feel the impact of cheap energy constraints well before others, the luckier countries may heed the warning and start to take meaningful actions. If so, it will be very important to know which countries will be the unlucky ones, and which may prove heedless of the new reality—and if possible, try not to be living in one of them.

The nature of collapse will feed into the urgency and scale of the defensive actions required. Paradoxically, a slow and grinding collapse may lull everyone into passivity, as each downward drop in living conditions will not be enough to offset the immediate pain of the long-term beneficial steps required. Such a process has been put forward as the reason that so few individuals tried to escape on their journey to the gas chambers during the Second World War. Each individual step was not enough of a jarring change to trigger the act of fight or flight that risked immediate death. At each stage there was always the hope that things would not turn out to be the worst. An immediate and obvious threat of death may have elicited a different reaction; in the same way, a sudden drop from relative comfort to penury may drive a greater and more urgent response.

5.2 Reducing Exposure to Fictitious Assets

If many financial assets may eventually turn out to be mostly fictitious, what can people do to protect their wealth? At some point, a sizeable percentage of the population will start asking this question, but by then it will be too late to act. Once a sizeable group starts attempting to unload financial assets, the weight of the selling will very rapidly drive down their worth and cause rapid inflation in the assets to which the money is flowing. The promised protection of a "diversified" portfolio of financial assets will also prove to be illusory as financial assets tend to move in synch [1], especially during market downturns, as a general market panic and liquidation takes hold. The vast majority will be left with greatly reduced financial asset values and rapidly escalating values of the alternative "real" assets. What I mean by "real" assets is those that provide real output independent of the financial markets, such as agricultural land, water, energy, and commodity supplies. The "smart money" has in some cases already started to move into these areas, as shown by the increasing limited partnerships focused on land investment and renewable energy sources, together with the purchasing of large tracts of productive land by governments and corporations. The increasing growth of farmland leasing and purchases in South America, Africa, and Canada is an example of investment funds, corporations and state entities moving into farmland, and out of financial assets.

There are ways of investing through the financial markets in commodities and in companies benefitting from constrained energy supplies. Renewable energy companies, energy supply companies (as more and more equipment and supplies are required for fossil-fuel energy exploration and extraction), and those providing energy efficiencies are good examples. Even then, the success of such investments will depend upon continued growth in those areas and the ongoing functioning of the financial markets. Unfortunately, much financial advice depends upon the fallacious assumption that "everything else will continue to function as normal" when in such an interconnected and financialized world that is the last thing that will happen. Having at least some investments whose value does not depend upon the ongoing functioning of financial markets will be a significant plus in your ability to deal with future challenges. Direct investments in local companies offering alternative energy and energy efficiency solutions can provide a way of bypassing the financial markets.

Some of the hardest decisions will be those pertaining to defined-benefits pensions and annuities. Do you believe that the institutions committing to what may be decades of future payments will be able to meet those commitments—or will they even exist in the future? If you are a pensioner of a bankrupt US municipality the reality may already be painfully obvious. However, if the organization promising to pay that pension is currently solvent then the decision becomes a much harder one; do you stay with that pension or do you turn it into a lump sum if that option is available to you? Also, what if the country you are in will be an energy exporter for many years, thus benefitting from energy shortages? The answer will be dependent upon your individual circumstances.

For many, such issues are a true predicament, which the dictionary defines as "an unpleasant situation which is difficult, or impossible, to extract oneself from". There will be no truly right answers, but we can take measures to buffer ourselves against the foreseeable impacts of the coming end of cheap energy. One fact that may help is that when interest rates are low the current value of those future payments is higher, thus, the lump sum will be greater. As interest rates rise, that pension lump sum becomes less. As energy supplies become constrained, and more and more people start to exit debt instruments, what do you think will happen to interest rates? The mixture of a loose monetary policy and constrained energy supplies will lead to rapidly increasing inflation and interest rates, together with repeated recessions—a mixture referred to as "stagflation" in the 1970s. With governments hell-bent on "printing" money to support the current dysfunctional status quo as long as possible, a rerun of stagflation is quite probable especially in the energy importing countries. The best place to invest your money back then was in real assets and precious metals. Getting into debt was also beneficial in that period, but this time may be very different in that respect. With much bigger drops in economic output, and constrained monetary income, debt may become unserviceable. In that case, through default and bankruptcy, your prized assets would have to be given up to the banks which may continue to enjoy the support and largesse of governments. This is the case with Greece, which is being pushed to sell off publicly owned companies, buildings and land to help pay off its debts in what has been referred to as a "US $71 billion fire-sale" [2].

The other area of possible dependency upon fictitious assets is from your job. Is your employment dependent upon the continued functioning of the debt and equity markets, mortgage lending, banking in general? It could also be dependent upon continued access to cheap energy. Jobs requiring long car-based commutes or those dependent on most customers arriving in a fleet of personal autos are certainly under threat. If your current job is well remunerated you may want to keep it though and use some of that income to fund alternative arrangements, including training in skills that will be useful in the future. If you are relatively young, studying to be a petroleum engineer may be a great career choice given that the average age of such engineers in the US oil and gas industry is around 50 [3]. Studying to provide financial services may not be so smart.

5.3 Location, Location, Location

As I noted in an earlier chapter, the shortage of cheap energy will affect different countries in different ways and in different timeframes.

- Energy exporters will be buffered by being able to charge more for their energy exports, and keeping back energy for their own use
- The middle-income developing countries, such as China and India, will be able to outbid the more advanced countries for marginal energy supplies. This is because they will still be working through the higher pay-off uses of energy, together with having much lower labor costs

The worst places to be will be the higher-income energy importing countries, such as the United States, Japan, and much of Europe, with only marginal returns on energy investments, or energy-importing countries with scant foreign exchange earnings with which to pay for imported energy.

The hardest hit of the latter will be the "ex"-energy exporters where, through increased domestic demand and/or reduced production, a country has moved from being an energy exporter to being an energy importer. Such countries tend not to have a well-developed industrial export sector; once they turn into an energy importer their economy may fall into a long-term decline. A poster child for this scenario is Egypt which recently turned into an energy importer, and is now dependent upon handouts from other countries. Any dreams of a new dawn in that country have run into the wall of decreasing energy availability, and a general inability to pay for imports of anything. In that situation, political unrest was inevitable. Other countries where the energy-export surplus is diminishing include Bahrain, Syria, and Yemen. Syria is embroiled in a civil war and the other two countries will have significant problems, but given their small scale and with other complicating, causative factors, their problems may not trigger a reassessment by other countries. Mexico has managed to stabilize its oil output in the past few years after another moderate decline, and has remained an energy exporter. The previous decline had already reduced exports by a third, so any further falls in output could put the country in danger of becoming an inconvenient version of Egypt on the US border [4, 5]. Those planning a happy retirement to countries in Central America may also want to check on the probable stability of their chosen country if faced with a constrained-energy future.

The example of the UK, a very stable, relatively rich country should have a sobering effect on others. After 200 years, its coal industry is nearly gone. Now its North Sea oil and gas fields are well into the depletion phase [6] with ongoing reductions of about 6 % a year in domestic oil and gas production [7]. And all but one of its nuclear power plants is scheduled to close within the next decade [8]. Despite the looming shortfalls, plans for a large program of new nuclear power stations have been scaled down and the timetable extended [9]. This will only increase the reliance upon natural gas-fired power stations, exacerbating the increase in energy imports. Under the European Union's (EU) 2009 Renewable Energy Directive, the UK has committed to having 15 % of its energy provided by renewable sources (wind, solar, hydro-electric, tidal, geo-thermal, and biomass) by 2020, from 3.3 % in 2010 [10]. Recently though, the UK government's 2020 projections for offshore wind capacity have been reduced by a third and the guaranteed price for offshore wind reduced as part of general government budget reductions [11]. The remaining hope is for shale gas production, but conditions are very different from those in the USA. The geology is more complex, the shale deposits are under heavily populated and environmentally sensitive areas, and the UK does not currently have the large numbers of horizontal drilling rigs required [12]. Short of ongoing recession to reduce energy usage, the UK faces severe challenges over the next decade which will provide an instructive dress rehearsal for the less energy-challenged countries. Further lessons may result as the very large financial sector within the UK economy, together with high levels of debt, exacerbates the effects of constrained energy supplies.

Most of the other EU countries are at different points in the energy constraint journey that the UK is traversing. The EU coal mining sector has already been severely reduced, with only Germany and Poland being significant coal producers. Increasing air quality regulations together with reductions in coal industry subsidies will drive further contraction. The only EU country with significant natural gas reserves, Holland, is also well into the depletion process with output falling by about 5 % per year [7]. In many EU countries, recent, short-sighted, austerity measures have created budgetary pressures leading to reductions in renewable energy subsidies. Hopes around shale gas will probably share the delays and shortfalls that will be experienced in the UK. The ongoing recessions and depressions across Europe have helped to delay confronting the energy supply issues in the short term, as reduced economic output, together with some ongoing efficiency gains, resulted in lower energy use. High population densities and relatively short distances between population centers also support energy efficient transportation such as trains and subways. Japan, too, shares much of the same dilemma that the EU has, with the shutdown of its nuclear power stations after Fukushima making things worse.

So, if Europe and Japan are not generally a good place to be as cheap energy supplies become more and more constrained, where would be a good place? Russia has the world's largest natural gas reserves, second largest coal reserves, and the ninth-largest crude oil reserves [13]. If any country will benefit from constraints on cheap energy it will be Russia, as it is able to charge higher and higher prices for its energy exports while still being able to supply its own economy. However, the autocratic and unpredictable style of the current government may reduce its allure. Australia stands out as a very good candidate. While its oil imports are rising as production falls, it has extensive natural gas resources that it has only just started to exploit [7] and it stands as the second largest coal exporter in the world with significant remaining reserves [14]. Norway also has very significant remaining oil and gas reserves, although its investments of previous oil revenues in financial assets abroad may have been a costly mistake. Sweden is an interesting candidate as it already meets more than a third of its energy consumption from renewable sources, and has a very strong commitment to becoming fossil-fuel free through energy efficiency and new renewable energy developments. Iceland has utilized its extensive geothermal resources to replace its dependency on fossil fuels, but with a population of only 350,000, a wickedly difficult language to learn, and all-day darkness for a significant part of the year it may not be for everyone.

In North America, Canada has a lot of pluses with extensive fossil fuel and renewable energy (especially hydroelectric) resources. The USA provides a complex problem. It is a major energy producer, but its prodigious energy consumption still necessitates significant energy imports. Fracking has provided a boost to oil and gas output, but this barely offsets the ongoing depletion of existing production. In short, the USA may prove that fracking may have much less long-term promise than is generally assumed. Worse, in the second half of the twentieth century, huge investments were directed into car-dependent suburbs while public transport was allowed to deteriorate or simply disappear. Some have hypothesized that its sheer energy inefficiency may provide the USA with a great scope for reductions in energy use, but any conservation may be greatly constrained by its previous investment choices. The older cities of the USA which tend to be much more compact and have better public transport infrastructures, such as New York, Boston, and Chicago, may have a great advantage in this respect. However, to take advantage of the cities' strengths, the vast American suburbs may have to be abandoned.

South America generally does seem to be in a relatively good place with supplies of fossil fuels, land for biofuels and food, and other raw materials. Brazil has a very high ratio of arable land to population, allowing for extensive growth of biofuel crops. It also has large offshore oil and gas reserves, which are only now being exploited, together with significant hydroelectric energy resources [6]. Ecuador, Bolivia, and Peru have significant oil and/or gas reserves, and Colombia has a lot of potential for hydro and solar energy [7]. Argentina lacks significant fossil fuel resources with respect to its consumption, but does have a low population density with respect to the amount of arable land available. Chile does not have significant fossil fuel resources, but does have very good hydroelectric potential and significant mineral resources which could be traded in exchange for energy imports. Venezuela is of course a major energy exporter [7] with a modest population and fertile lands, but may be open to extreme pressure from the USA if shale oil and gas prove to be a short-term chimera and Mexican oil output starts to decline again.

The "our oil under their ground" view that some of the very powerful countries seem to have of other country's natural resources, such as Venezuela, is an important factor that needs to be taken into account. Such resource-rich countries could turn into war zones, or be destabilized by outside forces. When energy constraints start impacting major countries, their first reaction may be to try to seize continued access to others' energy resources rather than make the required hard domestic choices. From this viewpoint, the Middle Eastern, African, and Central Asian energy exporters are not very good options. Finally, it goes without saying, any countries that are already politically unstable would be additionally unattractive.

5.4 Outlook for Change: Cognitive Dissonance, Vested Interests, and Inertia

Over more than 2 centuries of cheap energy and rapid growth a deep set of beliefs about the inevitability of such growth has developed. Traditional neo-classical economics, upon which much government policy making depends, provides an entire intellectual structure that chooses to ignore any notion about energy constraints or limits to growth. It is absolute economic dogma in that school of thought that the wonders of human ingenuity will always find a way around any limits. One wonders, if human ingenuity is so incredible, then why has rapid economic growth only existed since the use of cheap and abundant fossil fuels became widespread? In fact, that "human ingenuity" has needed ever increasing amounts of cheap energy to keep "solving" problems. The first trains would not have been much use without coal, nor the internal combustion engine without a liquid fuel. Unfortunately, the near-religious faith in human ingenuity and endless growth is deeply embedded within modern industrial societies. Individuals and groups tend to cling onto such deeply held beliefs even in the face of direct evidence to the contrary. People are driven to reduce the conflict between different beliefs they hold, or between their beliefs and new facts, in a process called cognitive dissonance [15]. Instead of doing what would seem logical, such as changing beliefs to fit new contrary information, people tend to twist, reject or diminish contrary information if it conflicts with their deeply held beliefs. In extreme cases, individuals not only cling to certain beliefs even though all evidence supporting them has disappeared [16], but they also increase their commitment to a belief in the face of contradictory facts [17]. People will also tend to seek out information that apparently backs up their beliefs [18] and discount or ignore anything that does not. Those with liberal political attitudes will probably not be regular viewers of Fox News while Tea Party supporters avoid MSNBC like the plague. The more a change in beliefs will lead to very difficult and discomfiting decisions, or threaten a person's current self-image, the more all these affects will be magnified.

An acceptance of limits to growth must lead to many painful decisions, such as an acceptance that much of our current wealth is illusory as it is based upon assumptions of future growth. Without this wealth comes an acceptance that many major business organizations and occupations are nonviable and that visions of a work-free retirement are probably illusory too. Also, without that extra bit of output each year, solving of social problems and inequalities will be much harder if not impossible. If future growth will not help the poorer members of society improve their lives, calls for redistribution from the better off to the less well off may become overwhelming. Once a lack of growth is accepted the implications can become extremely disconcerting both at the personal and the societal level. What politician would want to stand for election and tell this message to the voters, what chance would he or she have of ever being elected? Being one of the first to reject a widely held belief can also be extremely uncomfortable, exposing one to ridicule, verbal, and even physical aggression. This is especially so if others finds such contrary views to be very threatening to their current beliefs, self-image, and way of life. The pressure to conform can be incredibly hard to resist. This pressure may soon lead to increased deception on the part of ruling elites and attempts to divert attention from the harsh realities looming.

Fundamentally, human decision making is not a fully rational process. From the early twentieth century onward, businesses and politicians have become very skillful at taking advantage of the nonrational aspects of human decision-making [19]. In fact, the profession of public relations was founded by Edward Bernays [20], the nephew of Sigmund Freud. Now some of the biggest employers of psychologists and socio-psychologists are marketing and public relations groups laboring to change people's beliefs, to support ongoing self-deception, and to get people to buy products they did not previously know that they wanted. As people have generally become more overweight, clothes sizes magically get bigger and customers remain happily self-deceived. Happy customers buy more than unhappy ones. The fast food industry resists publishing the amount of calories in their food and drinks to support such self-deception. (Ever seen the person ordering the skinny latte with the extra whipped cream?) At the top of this pyramid are the private communications companies that dominate the news media. To help the paying advertisers, their broadcasts stay away from depressing stories and always seem to feature "feel good" stories at the end of the news programs.

The impact of these manipulative skills has been on display over the past decades in the attempts of the tobacco companies to forestall attempts to limit cigarette smoking by creating a perception of scientific uncertainty and challenging the integrity of their opponents. Such actions provide individuals with excuses to hold to their previous beliefs, and obfuscate the contradictory facts. Much the same has been done by corporate opponents of actions on climate change. The senior executives of major companies will feel great pressure to maintain the fiction of the future growth that supports their shareholder's wealth—and their own. Many leaders may also be simply caught up in their own cognitive dissonance and inertia, as with the International Energy Association that keeps producing optimistic reports of future energy supplies which, as regular as clockwork, are shown to be erroneous (if not downright deceptive) within a year or two.

The end result of these inertial forces can be a society unable to react rationally to major threats. Unfortunate facts will be avoided, and anything providing support to the current beliefs will be sought out. Shale gas and windmills will save us all; climate change is rubbish because it snowed so much recently; economic growth will continue forever! Historically, these inertial forces have been overcome when sudden, immediate threats present themselves and galvanize society in response. Pearl Harbor, the 9/11 attack, and Argentina's invasion of the Falkland Islands are just some examples. Car factories can be swiftly converted to produce planes, or armadas can be constituted and dispatched incredibly rapidly after such galvanizing events. Unfortunately, the end of cheap energy will not present itself as one sudden crisis. It will creep up, more like an ongoing decline interspersed with smaller crises. Society will incrementally collapse accepting each lower standard of living as a small change from the past. As Diamond [21] has pointed out, for the average citizen the collapse of civilizations, such as the Maya and Roman, may have been imperceptible as it was experienced as a series of crises and stabilizations stemming a number of generations.

Without a galvanizing event inertia may be overcome, but usually a long period of accumulating contrary facts is required. After more than two decades of talking about the need for action, Western societies are still just talking about making small changes to limit the impact of climate change. Meanwhile, discussions about peak oil almost invariably focus on finding a new fuel for the internal combustion engine—and not on the obvious need for its demise. Given the scale of the uncomfortable changes needed, and the depths of beliefs about human ingenuity driving growth forever, it will take a long time for the accumulation of contrary indicators to drive an acceptance of the need for change. Even then, only token responses may be forthcoming for a while. As the impact of scarce cheap energy will hit different countries at different speeds, it is possible that the visions of collapse in some countries may galvanize the luckier countries into actions to stop themselves from meeting the same fate.

5.5 Nature of Collapse

For a person walking through fog towards the edge of a cliff only the last step taken will produce a bad outcome. Each step before that will not change his health or circumstances, but with that last step everything changes. Even if he survived the fall, he will be lying at the bottom of the cliff with broken bones and other injuries and may never be able to return to his previous state. If the tide is coming in, remaining alive may not be possible. The reaction of the financial system to change is very much like walking towards the edge of a cliff; before reaching certain limits (the edge of the cliff) things hardly change. But once those limits have been breached, things can change extremely rapidly and a return to previous circumstances may not be possible. Such events are known as crashes, which often seem to come from nowhere. Many may have known that things were becoming precarious, but none could predict when the financial system would fall off the cliff.

Scientists call this phenomenon "bounded resilience"; where a given state is only possible within limiting boundaries. The human body can only survive in a small band of temperatures, and it uses offsetting mechanisms to try to stay within that range. When the body approaches its low temperature threshold shivering sets in, when approaching the high temperature threshold and sweating is triggered. Such mechanisms are called *negative feedbacks*, designed to offset a move beyond a boundary. The human body also utilizes mechanisms to intensify a process, for example, blood clotting in response to a cut. The damaged skin releases chemicals that activate blood platelets, and the activated platelets release chemicals to activate more platelets. This reinforcing cycle causes a rapid response that delivers enough platelets to form a blood clot. Such a self-reinforcing process is known as a *positive feedback*. Complex systems such as the human body, the financial system, and the global economy tend to exhibit negative feedback mechanisms, but when these fail, positive feedback can take over and rapidly move the system to a different state. Once a selling panic takes over the financial markets selling tends to create more selling, as more and more people try to get out before they lose everything. This is exacerbated by the forced selling (known as "margin calls") for those who have purchased assets using borrowed money.

Since 2008, the central banks of the world have been pouring trillions of dollars, euros, yen, pounds etc. into the global financial system, and taking many other steps to keep it within its bounds of sustainability. Without this huge, ongoing support the system would have collapsed, taking the global economy with it. This is akin to continuously feeding blood into a patient with a crushed leg which needs to be amputated to stop the bleeding. The patient may be kept alive but their general condition will continue to worsen until the leg is removed and the bleeding stopped. Without the removal of the overgrown financial system, together with growing, unsustainable debt levels, the global financial system cannot be removed from its monetary I.V. feed. Instead it will continue to weaken and place its hosts—our economies and societies—at greater and greater risk. If it does not collapse beforehand, cheap energy constraints will provide the knockout blow to the global financial system, forcing it outside the sustainable bounds and triggering the positive feedbacks that will drive self-reinforcing collapse. A replay of the 2008–2009 financial collapse will take place, but on a larger scale and without meaningful recovery.

Some have postulated that such a collapse will provide the seeds of a partial recovery, as the reduction in energy demand will create a short-term surplus of supply. The suddenly plentiful supply of cheap energy will then support a recovery or it will, until demand soaks up the surplus and a new collapse ensues. As the supply of cheap energy declines over time there will be lower peaks and deeper troughs, as the economy declines through repeated collapses and recoveries. This scenario may be a little optimistic though as it fails to take into account the extreme difficulty of rebuilding the societal and economic complexity that will be destroyed during the collapse phase. Replacing such things as bankrupt companies, disrupted trade networks, and the trust inherent in financial relationships can be a very long process, as seen after the 1930s crash. The need to reinvent ways of doing things with increasingly costly energy will also reduce the pace of recoveries. So, the peaks may be lower and take longer to attain, and the resulting troughs deeper also. Such crises will also inhibit possible alternatives which require complex coordination and up front capital investment. After the failure of trade and financial networks, how will the myriad materials and components required to build a wind-turbine be brought together? What happens if a number of the intermediate producers went bankrupt during the crisis? Given the general behaviour of complex systems, and the historical examples of financial crashes, it is safe to say that the crash(es) will be relatively sudden and deep. In between there may be periods of relative stability with some growth, but any recovery will hit biophysical limits (the supply of cheap energy, the over-carbonated atmosphere) as well as being constrained by the great difficulty of resurrecting complex social and economic networks in an era of expensive energy.

5.6 Summary

I have tried to cover the main factors that will drive personal choices with respect to the coming cheap energy crisis. Given the complexities of the world in which we live, no one can accurately predict the absolute timing and detailed impacts of this crisis. Also, individual circumstances will be very different; some will be rich and others not so rich; some will be young and some will be older; some will have more responsibilities than others; countries of residence and citizenship will vary greatly. None of us will be spared from the impacts of the coming crisis; all we can do is try to reduce its impact upon us by taking the defensive steps that are open to us. I hope that this book has helped you in understanding the future challenges and in formulating what your own steps need to be.

References

1. Authers J (2010) The fearful rise of markets: global bubbles, synchronized meltdowns, and how to prevent them in the future. FT Press
2. Taylor A (2013) The assets Greece needs to sell to get another bailout. Business Insider. http://www.businessinsider.com/greek-assets-sale-2011-5?op=1. Accessed 5 August 2013
3. Kemp J (2012) Have you considered a career in petro-engineering? Reuters. http://www.reuters.com/article/2012/01/18/column-oil-engineers-idUSL6E8CD43S20120118. Accessed 25 July 2013
4. N/A (2013) Mexico week: lower Mexican oil production contributes to lower crude oil exports to U.S. U.S. Energy Information Administration. http://www.eia.gov/todayinenergy/detail.cfm?id=11251. Accessed 5 August 2013

5. N/A (2013) Update 1-Mexico oil output slumps to near two-year low in May. Reuters. http://www.reuters.com/article/2013/06/28/mexico-oil-idUSL2N0F41RK20130628. Accessed 25 July 2013
6. N/A (2013) United Kingdom country data, U.S. Energy information administration. http://www.eia.gov/countries/country-data.cfm?fips=uk. Accessed 27 July 2013
7. Campbell C (2013) Campbell's Atlas of oil and gas depletion. Springer
8. Wintour P, Inman P (2013), Nuclear power gets £ 10 bn financial guarantee boost, the guardian. http://www.guardian.co.uk/environment/2013/jun/27/nuclear-power-10billion-financial-guarantee. Accessed 29 July 2013
9. Gosden E (2013) Government extends new nuclear power station timetable by five years, confirms first plant will cost up to £ 14 bn, the telegraph. http://www.telegraph.co.uk/finance/newsbysector/energy/9954515/Government-extends-new-nuclear-power-station-timetable-by-five-years.html. Accessed 6 August 2013
10. N/A (2013) Report from the commission to the European parliament, the council, the European economic and social committee and the committee of, the regions. The European commission. http://eur-lex.europa.eu/LexUriServ/LexUriServ.do?uri=CELEX:52013DC0175:EN:NOT. Accessed 6 August 2013
11. Harvey F (2013) Future of UK offshore wind power in 'serious doubt'. The Guardian. http://www.guardian.co.uk/environment/2013/jul/09/future-uk-offshore-wind-power. Accessed 8 August 2013
12. White G, Rowley E (2013) Forget a quick shale gas revolution here—we don't have the technology. The Telegraph. http://www.telegraph.co.uk/finance/commodities/10151313/Forget-a-quick-shale-gas-revolution-here-we-dont-have-the-technology.html. Accessed 15 August 2013
13. N/A (2013) Russia country data. U.S. Energy Information Administration. http://www.eia.gov/countries/country-data.cfm. Accessed 8 August 2013
14. N/A (2013) Coal statistics. World Coal Association. http://www.worldcoal.org/resources/coal-statistics/. Accessed 8 August 2013
15. Festinger L et al (1956) When prophecy fails. Harper-Torchbooks, USA
16. Brockner J (Jan 1992) The escalation of commitment to a failing course of action: toward theoretical progress. Acad Manage Rev 17(1):39–61
17. Baumeister RF, Kathleen D (2007) Encyclopedia of Social Psychology. SAGE Publications
18. Nickerson RS (1998) Confirmation bias: a ubiquitous phenomenon in many guises. Rev Gen Psychol 2(2):175–220
19. Curtis A (2002) The century of the self. British Broadcasting Company
20. Bernays E (1928) Propaganda. Horace Liveright
21. Diamond J (2004) Collapse: how societies decide to fail or succeed. Viking Adult

Index

R. Boyd, *Energy and the Financial System,* SpringerBriefs in Energy,
DOI 10.1007/978-3-319-04238-1, © Roger Boyd 2013

FSC
www.fsc.org
MIX
Papier | Fördert gute Waldnutzung
FSC® C083411

Kolibri 360 GmbH
Ferdinand-Jühlke-Straße 7
99095 Erfurt, Deutschland
produktsicherheit@kolibri360.de